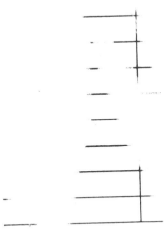

REUSE TECHNIQUES FOR VLSI DESIGN

REUSE TECHNIQUES FOR VLSI DESIGN

Edited by

Ralf Seepold

Forschungszentrum Informatik (FZI), Karlsruhe

and

Arno Kunzmann

*Forschungszentrum Informatik (FZI), Karlsruhe
and Internext GmbH, Karlsruhe*

KLUWER ACADEMIC PUBLISHERS
BOSTON / DORDRECHT / LONDON

A C.I.P. Catalogue record for this book is available from the Library of Congress.

ISBN 0-7923-8476-8

Published by Kluwer Academic Publishers,
P.O. Box 17, 3300 AA Dordrecht, The Netherlands

Sold and distributed in North, Central and South America
by Kluwer Academic Publishers,
101 Philip Drive, Norwell, MA 02061, U.S.A.

In all other countries, sold and distributed
by Kluwer Academic Publishers,
P.O. Box 322, 3300 AH Dordrecht, The Netherlands

Printed on acid-free paper

All Rights Reserved
©1999 Kluwer Academic Publishers
No part of the material protected by this copyright notice may be reproduced or
utilized in any form or by any means, electronic or mechanical,
including photocopying, recording or by any information storage and
retrieval system, without written permission from the copyright owner.

Printed in the Netherlands

CONTENTS

FIGURES		IX
TABLES		XI
PREFACE		XIII

1 ECSI, VSIA AND MEDEA - HOW INTERNATIONAL ORGANISATIONS SUPPORT REUSABILITY   1
A. Sauer
   1.1 Abstract   1
   1.2 The System-on-a-chip Challenge   2
   1.3 ECSI - The European CAD Standardisation Initiative   2
   1.4 VSIA - The Virtual Socket Interface Alliance   4
   1.5 MEDEA - The Eureka Project on Micro-Electronic Development for European Applications   6

2 ANALYZING THE COST OF DESIGN FOR REUSE   9
I. Moussa, M. Diaz Nava and A. A. Jerraya
   2.1 Abstract   9
   2.2 Introduction   10
   2.3 Case Study: ATM Shaper Design   10
   2.4 Specific and Reusable Blocks   13
   2.5 Comparing Reusable and Specific Components   16
   2.6 Conclusion   19

3 A FLEXIBLE CLASSIFICATION MODEL FOR REUSE OF VIRTUAL COMPONENTS   21
N. Faulhaber and R. Seepold
   3.1 Introduction   21
   3.2 Objective   22
   3.3 State of the Art   22
   3.4 RMS Similarity Metric   24
   3.5 The RMS-Taxonomy   27
   3.6 Extended RMS-Classification   29
   3.7 Implementation   31
   3.8 Application of the Model   32
   3.9 Conclusion and Outlook   35

4 AN INTEGRATED APPROACH TOWARDS A CORPORATE DESIGN REUSE STRATEGY 37
U. Schlichtmann and B. Wurth
4.1	Introduction	37
4.2	Why is Design Reuse Difficult?	39
4.3	Core Supply Process	40
4.4	Organization	43
4.5	Business Models	44
4.6	Ensuring Core Quality	45
4.7	Technical Issues	45
4.8	The Overall Strategy	47

5 DESIGN METHODOLOGY FOR IP PROVIDERS 49
J. Haase, T. Oberthür and M. Oberwestberg
5.1	How to Become an IP Vendor	49
5.2	IP Database Structure	50
5.3	Documentation of IP	51
5.4	Simulation Testbench Philosophy	53
5.5	Release Management	56
5.6	Dual Language DesignObjects	56
5.7	Scalable DesignObjects	58
5.8	Experience from Reuse Projects	59
5.9	Conclusions	61

6 HARD IP REUSE METHODOLOGY FOR EMBEDDED CORES 63
W. Eisenmann, S. Scharfenberg, D. Seidler,
J. Geishauser, H. Ranise and P. Schindler
6.1	Introduction	63
6.2	Simulation Model Generation	64
6.3	StarterKit Simulation Environment	65
6.4	Timing Characterization and Timing Models	67
6.5	Frontend Views and Embedded Core Test Methodologies	71
6.6	Backend Views and Backend Design	73
6.7	IP Repository	75
6.8	Next Steps	77

7 A REUSE LIBRARY APPROACH IN ENGINEERING CONTEXT 79
S. Müller
7.1	Motivation	79
7.2	Project Description and Objectives	80
7.3	Reuse Methodology	81
7.4	Module Administration	84

7.5	Presentation and Access to Modules	86
7.6	Measurement	88
7.7	Summary	88
7.8	Conclusion	89
7.9	Status of the Work	89
7.10	Outlook	89

8 ASPECTS OF REUSE IN THE DESIGN OF MIXED-SIGNAL SYSTEMS   91
F. Heuschen, Ch. Grimm and K. Waldschmidt

8.1	Abstract	91
8.2	Introduction	92
8.3	Top-Down Design Flow	92
8.4	Databases and Reuse	94
8.5	Summary	101

9 DESIGN REUSE EXPERIMENT FOR ANALOG MODULES "DREAM"   103
V. Meyer zu Bexten and A. Stürmer

9.1	Introduction	103
9.2	Requirements for Reuse of Analog Blocks	104
9.3	Implementation	105
9.4	Experience	107
9.5	Future work	108
9.6	Acknowledgement	109

10 REDESIGN OF AN MPEG-2-HDTV VIDEO DECODER CONSIDERING REUSE ASPECTS   111
H.-J. Brand, R. Siegmund, St. Riedel,
K. Hesse, and D. Müller

10.1	Introduction	111
10.2	Design Reuse	112
10.3	Redesign of an MPEG-2-HDTV Video Decoder	116
10.4	Design by Reuse (Inverse Quantiser)	118
10.5	Design for Reuse (IDCT - Inverse Discrete Cosine Transform)	119
10.6	Summary	122

11 REUSE CONCEPTS IN GROPIUS   125
D. Eisenbiegler and C. Blumenröhr

11.1	Abstract	125
11.2	Introduction	126
11.3	Gropius - a Survey	127
11.4	Design Reuse across Abstraction Levels	127
11.5	Everything can be Abbreviated	131

	11.6	Polymorphism	131
	11.7	Parameterization with Circuits	132
	11.8	Regularity	132
	11.9	Strict Separation between Functional and Temporal Aspects	132
	11.10	Uniform Communication Protocol at the System Level	135
	11.11	Conclusion	136

12 LEGAL ASPECTS OF REUSE OF INTELLECTUAL PROPERTY   139
R. Vogel

	12.1	Issues	139
	12.2	Legal Situation	140
	12.3	Contractual and Technical Remedies	142

13 REFERENCES   145

INDEX   151

FIGURES

Figure 2.1	Shaper architecture	11
Figure 2.2	Shaper partitioning	13
Figure 2.3	Partitioning and modular design techniques	14
Figure 2.4	Cost of design versus reuse	19
Figure 3.1	Evaluation of retrieval mechanisms	23
Figure 3.2	RMS similarity metric	25
Figure 3.3	Conceptual similarities	26
Figure 3.4	Example of asymmetric similarities	26
Figure 3.5	DAG and symbol of a VCA	27
Figure 3.6	RMS-Taxonomy	28
Figure 3.7	Extended RMS-Classification	30
Figure 3.8	Retrieval algorithm	31
Figure 3.9	ORACLE web architecture	32
Figure 3.10	Specification of the ALU	34
Figure 3.11	Attributes of ALU_1	35
Figure 4.1	Core supply process	40
Figure 4.2	Organisation	43
Figure 5.1	Example of a directory structure	51
Figure 5.2	Top level testbench generator window	53
Figure 5.3	General testbench structure	55
Figure 5.4	Example of translation flow	57
Figure 5.5	Testbench environment	60
Figure 6.1	Hard IP deliverables	64
Figure 6.2	VMC model generation and simulator build	65
Figure 6.3	Modelling layers	66
Figure 6.4	Reference structure design	66
Figure 6.5	Data flow for a software driven simulation	67
Figure 6.6	Timing characterization flow	68
Figure 6.7	Black box timing groups	69
Figure 6.8	Multiplexing and internal boundary scan	71

Figure 6.9	Full Scan and built in Self Test	72
Figure 6.10	Full blockage and shrink wrap abstracts	74
Figure 6.11	IP repository architecture	76
Figure 7.1	Work groups and dependencies on results	82
Figure 7.2	Library submission flow	84
Figure 7.3	Example of the library contents list	85
Figure 7.4	Entry page of the reuse library	87
Figure 8.1	Design flow for mixed signal systems	93
Figure 8.2	Design flow extended by a reuse database	95
Figure 8.3	Internals of an operational amplifier	96
Figure 8.4	Template for Sallen-Key filter	97
Figure 8.5	Module description template excerpt	99
Figure 8.6	SPICE input file example	100
Figure 9.1	Structure of the DREAM implementation	106
Figure 9.2	DREAM search engine	107
Figure 9.3	Data with circuit schematic preview	108
Figure 9.4	Data with design reentry window	109
Figure 9.5	Online entry/editing of a design specification	110
Figure 10.1	HiPEG architecture	116
Figure 10.2	Interface-based design	121
Figure 11.1	Syntax of Gropius	128
Figure 11.2	Gropius	130
Figure 11.3	Example I of derived structures	133
Figure 11.4	Example II of derived structures	134
Figure 11.5	Examples of communication schemes	136

TABLES

Table 1.1	Associate members speciality	3
Table 1.2	Activities	4
Table 1.3	VSI Alliance	5
Table 1.4	ECSI DRG proposal	6
Table 1.5	MEDEA cooperation	7
Table 1.6	MEDEA core competencies	8
Table 2.1	Cost of design	17
Table 2.2	Design productivity	18
Table 3.1	VCAs of the CV Function	33
Table 3.2	CAs and VCAs for input/output	33
Table 5.1	Gate count for decoder family	61
Table 5.2	Effort for decoder development	61
Table 6.1	Number of required characterization runs	70
Table 7.1	Comment text	86
Table 8.1	Technology table field definition	98
Table 8.2	Technology table example	98
Table 10.1	IP component classification	113
Table 10.2	IDCT features	119
Table 10.3	IDCT parameter set	120
Table 11.1	Basic constructs of Gropius	129

PREFACE

The design of microelectronic systems is strongly influenced by the fact that transistor and feature size have continuously decreased, while density and frequency have increased. The development of new technologies has supported the achieved gain by providing new mechanisms and equipment to improve design. In a first phase, design flows and frameworks had been regarded as an adequate methodology to manage increasing complexity, which was a direct consequence of the technology gain.

Due to that fact that short time to market strongly influences the prosperity of a product, new and innovative methodologies have to be investigated to create future-oriented products associated to an aggressive time-scale. Therefore, in a second phase, reuse of components has been regarded as a key enabler to amplify the availability of both applicable high-end manufacturing technologies and system design tools to manage System-on-a-Chip (SoC) design.

The "design gap" that is referred to chip capacity versus design capabilities is going to widen out, if the development of EDA tools is restricted to improvements in well known domains of research or to the tuning of complex algorithms. A successful bridging of this gap can be achieved by the application of reuse-based design methodologies, since it is accepted that design reuse is a key technology and its paradigm can be compared to that of high-level synthesis.

Reuse of Intellectual Property (IP) and virtual components (VC) is required for SoC development. Since SoC design must face the requirements of today's consumer market, it is a fundamental effort to design common interface standards, to enforce accommodation of IP and to transfer IP in a virtual market place. In this market place currently under development, the participation of companies closely depends on the successful adoption/modification of a reuse-oriented business model and on the realization of a methodological shift from isolated reuse to intra-company reuse, and finally, to inter-company reuse. A major challenge there is that IP has to be categorized and the most common types of IP have to be presented by their distinctive characteristics.

The purpose of this book is to reflect the current state of the art in design reuse for microelectronic systems. To accomplish this goal, typical representatives from leading research and application areas have been selected to document mature approaches and their latest research results.

Firstly, it sets out the background and support from international organisations that enforce SoC design by reuse-oriented methodologies. This overview is followed by a number of technical presentations covering different requirements of the reuse domain. This is presented from different points of view, i.e., IP provider, IP user, designer, isolated reuse, intra-company or inter-company reuse. More general systems or case studies, e.g., metrics, are followed by comprehensive reuse systems, e.g., reuse management systems partly including business models.

Since design reuse must not be restricted to digital components, mixed-signal and analog reuse approaches are presented. In parallel to the digital domain, this area covers research in reuse database design.

Design verification and legal aspects are two important topics that are closely related to the realization of design reuse. These hot topics are covered by presentations that finalize the survey of outstanding research, development and application of design reuse for SoC design.

1 ECSI, VSIA AND MEDEA - HOW INTERNATIONAL ORGANISATIONS SUPPORT REUSABILITY

A. Sauer

Chairman of ECSI, Grenoble, France
Vice-Chairman of MEDEA, Paris, France

1.1 ABSTRACT

In the challenging field of designing Systems-on-a-Chip three international organisations support their customers by different activities: ECSI has broadened its activities, e.g. to the virtual component domain, and intensified, e.g. in the language domain, and has become a member of VSIA and its European representative. VSIA, founded in 1996, has installed 7 Development Working Groups for the development of interface standards and to support the reusability of virtual components. MEDEA supports the European institutes and industry in related projects by forcing horizontal co-operation between European system companies and IC manufacturers.

1.2 THE SYSTEM-ON-A-CHIP CHALLENGE

The observation that the number of transistors on a chip doubles every 18 months generally is referred to as Moore's Law. This development made it possible to shrink entire motherboards containing dozens of chips on to a single chip. Application specific systems-on-a-chip represent considerable costs savings when manufacturing high volume electronic devices. Systems-on-a-chip have been around since several years, but now with ever progressing IC manufacturing technologies and the shift from analogue to digital electronics, systems-on-a-chip with as many as 100 million transistors are predicted by Dataquest for the year 2000. Europe, especially in the area of mixed signal chips, represents according to Dataquest around 35% to 40% of the mixed signal market, but East Asian companies are coming up.

Designing and manufacturing systems-on-a-chip is one of the most challenging projects facing system houses. Combining different functions on to the same chip is economically only possible when sophisticated design tools are available to simulate the operation of a later chip at an early design stage, as the projects turn out to be extremely complex, expensive, time consuming and risky.

To overcome these problems a new paradigm has been created under which immense effort have been brought together: reuse of existing designs. International organisations support this challenge with their different possibilities: ECSI by its standardisation efforts in the general field of CAD, VSIA by its specific activities in IP standardisation and MEDEA by its collaborative projects between system and IC companies.

1.3 ECSI - THE EUROPEAN CAD STANDARDISATION INITIATIVE

ECSI was founded in 1993 in Grenoble, France, as the European platform for CAD standardisation. Sponsored in the first year by an Esprit project, ECSI counted in the end of 1997

- 15 industrial members,
- 8 associate members, and
- 40 individual members.

In its beginning ECSI was active in the fields of interfaces (EDIF), languages (VHDL) and frameworks (CFI). In order to intensify the work in these domains, 3 Technical Centers were connected to ECSI: University of Manchester (EDIF), IMT (VHDL) and GMD (CFI). This concept was very convincing and some more institutes joined ECSI with different focus (see Table 1.1).

ECSI has broadened its activities, e.g. to the virtual component domain, and intensified, e.g. in the language domain. ECSI has generated significant momentum in all concerned areas of activities through participation in the European projects, through the improved co-operation with ECSI members and partners and through the provision of active services to members as well as promotion and publication actions:

• University of Manchester
  – Information modelling/EDIF
• CNET de Grenoble
  – Telecommunication Systems Design
• University of Twente
  – VHDL based design methodology
• CEFRIEL / P. di Milano
  – Hardware/Software Co-design
• Universidad de Cantabria
	• ITE, Warszawa
	  – Semiconductor technology
	• Technische Universität, Wien
	  – ECSI WWW Home Page
	  – Embedded system design
	• EPFL de Lausanne
	  – Analog/Mixed-Signal design methodology and tools
	• University of Patras
	  – System Level Design

Table 1.1 Associate members speciality

ECSI was involved in 4 European projects, and 3 other project proposals have been submitted. The regular publication of the ECSI Letter is an important achievement among the awareness and publicity activities. Issues are regularly published on a quarterly basis. ECSI is present in many ways at international event, e.g. having a booth at DATE[1]. ECSI is organising several Workshops and seminars for the European industry.

During 1997, ECSI has become the VSI Alliance representative in Europe. ECSI is member of the OMF Executive Committee, EDAC, the EDA Roadmap Industry Council, the EIA-EDIF division, VHDL International. A summary of ECSI's activities is included in Table 1.2.

In order to support the European contribution in the field of virtual components and its reusability ECSI became a member of VSIA in 1997. Jean Mermet, the ECSI Director, published our intention already in the October 1996 issue of the ECSI Letter. ECSI mainly supports work of VSIA by contributing to the Development Working Groups, installing and maintaining Design Review Working Groups. 2 ECSI Offices, supporting VSIA activities in Europe, were created during 1998:

- FZI in Karlsruhe for the German speaking countries,
- Aspen Enterprise Ltd. for Northern Europe.

Two meetings were organised by ECSI in May and September in Karlsruhe, one meeting in London Heathrow in September.

---

1. Design Automation and Test in Europe (http://www.date-conference.com)

4   REUSE TECHNIQUES FOR VLSI DESIGN

> - Creation of a German and an English office
> - Participation in the standardisation work:
>   - Analog/mixed signal VHDL
>   - OMF
>   - SLDL
>   - VSI-A DWG's and DRG's
>   - EDIF/ IBIS/ Express/ ALF/ CHDStd/ Test
> - Dissemination/Awareness
>   - VHDL newsletter, ECSI letter, ...
>   - Workshops
>   - WWW: www.ecsi.org
>   - VSI DRG's and other meetings

Table 1.2   Activities

1.4 VSIA - THE VIRTUAL SOCKET INTERFACE ALLIANCE

The silicon capacity exceeds design implementation capability by far. Following Sematech, the "Design Gap" is significant and continues to grow:
- 58%/year growth in silicon capacity,
- 21%/year growth in design capability.

System on Chip (SOC) capability is a looming challenge, and an old but predictable story. It needs a paradigm shift in order to "catch up"
- Design reuse,
- Mix and match (IP exchange),
- New business models and infrastructure.

This was the reason why VSIA was created in 1996 (see Table 1.3). More than 200 companies joined until now. They represent all major sectors of electronic industry. There is still room for the participation of system companies, since they are the most beneficial users of VSIA reuse methodology.

VSI offers vision and enablers to achieve SOC by use of virtual components. It is most effective way to create innovative SOC with competitive time-to-market. Design reuse is the most effective way to implement efficiency.

VSI has establish its work in 7 DWGs:
- Analog and mixed signals,
- System level design,
- On-chip bus,

- Manufacturing related test,
- Implementation/verification,
- IP protection,
- Virtual component transfer (VCT).

> - The Organization
>   - Launched in September 1996 with 35 companies, VSIA now has over 150 members with excellent System House, Semiconductor, IP Provider and EDA/Service representation
> - The Vision
>   - To make system-chip design a practical reality by enabling the mix and match of component blocks (virtual components) from multiple sources onto a single silicon chip
> - The Mission
>   - To Establish and promote
>   - o The unifying vision for the system-chip industry
>   - o The open technical standards required to mix and match Intellectual Property (IP) from multiple sources.

Table 1.3 VSI Alliance

Besides the technical work carried out by the DWG's, management is performed by the Steering WG with specific committees devoted to solving the technical co-operation between DWG's (Technical Committee) and promotion of the VSI vision and results (Marketing Committee).

The organisational structure is supported by offices in USA, Japan and Europe. ECSI is the VSIA representative in Europe.

ECSI, together with VSIA, has introduced early 1998 the Design Review Working Groups (DRG) with the purpose of having broader participation by members in the specification development and review processes:

- Attracting more and better comments from members during review of new specifications
- Getting more expertise and diversity in the specification development and review processes
- Providing more and more clear benefits of membership in the alliance
- Stimulate broader and earlier adoption of new specifications.

The structure of the ECSI's contribution to VSIA is shown in Table 1.4.

Realisation
- Creation of three language based DRGs in Europe:
  English, French, German
- Creation of group of ECSI representatives:
  – the group of 5 independent experts (corresponding to the domain of activity of DWGs)

SLD	Jean Mermet (ECSI, TIMA)
On-Chip Bus	Tony Gore (ECSI, OMIMO)
Mfg. Rel. Test	Adam Osseiran (ECSI, EPFL)
AMS	Alain Vachoux (ECSI, TIMA)
IP Protection	Adam Morawiec (ECSI, TIMA)

  – the group will ensure the representation of European VSI members in DWG (participation in US meetings)
  – the group will serve as a liaison between VSIA and European VSI member companies
  – management of meeting organisation, contacts, e-mail groups
  – preparation of minutes, summaries from the actual advance-

---

Table 1.4   ECSI DRG proposal

1.5 MEDEA - THE EUREKA PROJECT ON MICRO-ELECTRONIC DEVELOPMENT FOR EUROPEAN APPLICATIONS

Microelectronics production is of considerable importance in Europe. It is vital and strategic for the electronics systems industry in Europe, which is by far larger and which is facing markets with global competition, high growth rates, shortest innovation cycles and permanent price decreases. Europe developed an encouraging position in IC wafer fabrication due to the JESSI programme. On the application side, Europe's system houses have been able to use Europe's improved semiconductor expertise to develop state-of-the-art chips that are the heart of several world beating electronic systems contributing to Europe's competitiveness. In order to keep this strength and improve it, MEDEA as one of the largest EUREKA programmes was started in January 1997 with a runtime of 4 years (see Table 1.5). Its objective is: "Strengthen the global competitiveness of the European Microelectronics Industry, through R&D co-operations in technology and leading applications".

Key of this R&D programme is the strong co-operation between system companies and IC companies in the most important application areas: Multimedia, Communications and Automotive Electronics. Projects in these application fields work on new algorithms/methods and prototype implementations in systems-on-chips using advanced silicon. The 'glue' of this work is the availability of advanced CAD and design tools, methodologies and libraries. Some of these necessities are not available from the big American CAD vendors. MEDEA therefore has introduced a CAD programme which currently contains 7 projects. These projects are running in the following key areas:

- **Deep Submicron Design:** The projects are on the way in this area dealing with the development of new tools for the design flows and methods for 0.25mm, 0.18mm technologies and below. This includes methodologies how to handle very complex circuits with a multilevel approach, how to develop and use hierarchical test methods for system debug, system maintainability and manufacturing tests. Tools for modelling and parameterisation of IC's in terms of physical coupling are part of the development.

- **System Level Design:** The projects in this area cover HW/SW co-design tools and methodologies, reusability of HW and SW modules. Hierarchical design flows in which IP's could be introduced are also part of the work in this area. Tools for IP's development and interface standardisation are part. The projects contribute to international standardisation activities, e.g. ECSI and VSIA.

- **Analogue Design:** The project in this area has the goal to increase the design productivity and quality of analogue and mixed-signal IC designs. A library of basic functions used by the tools and required for demonstrator applications will be developed. Reusability of analogue and mixed-signal modules is also part of the project.

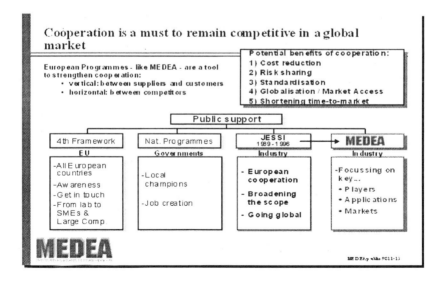

Table 1.5 MEDEA cooperation

- **RF-Design:** A project develops tools and methodologies for the development of RF building blocks required in future terminals in the fields of Digital European Cordless Telephone, Private Mobile Radio Communication and Global System for Mobile Communication.

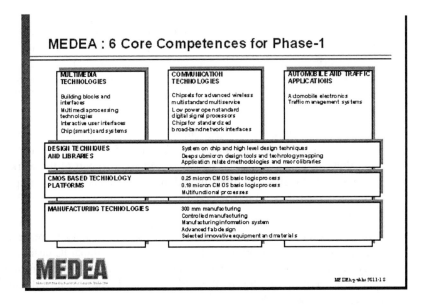

Table 1.6 MEDEA core competencies

This last mentioned project demonstrates the principle and the advantage of the MEDEA project structure: the vertical co-operation of projects. Application projects need results from MEDEA CAD projects and technology projects in order to achieve their goals, and vice versa: MEDEA CAD and technology projects need input from the application projects in order to define their own goals. To achieve these goals a broad variety of expertise is necessary: MEDEA includes in its CAD programme 25 companies, 9 institutes/universities from 6 European countries.

The MEDEA Design and CAD projects cover an essential part of the design part of the SIA Roadmap, esp. this part which is needed by the European system companies.

Strengthen the global competitiveness of the European Microelectronics Industry, through R&D co-operations in technology and leading applications

For further information visit the Web-sites:

- www.ecsi.org
- www.vsia.org
- www.medea.org

2 ANALYZING THE COST OF DESIGN FOR REUSE

I. Moussa[*], M. Diaz Nava[**] and A. A. Jerraya[*]

[*] TIMA lab., SLS Group Grenoble, France
[**] ST Microelectronics Crolles, France

2.1 ABSTRACT

This paper evaluates the cost of design for reuse through the design of an Asynchronous Transfer Mode (ATM) Shaper macro. This macro block is made of 2 million transistors and is composed of several modules or sub-systems. Some of these blocks are designed in order to be reused in other applications. This study shows that the modules designed for reuse require up to 2.5 more time than those specifically designed for our Shaper. The extra cost is mainly related to the time spent in three main areas: extensive analysis of the block and its potential application domains, the development of a more robust and complete test bench environment, and the preparation of good documentation. Furthermore, we will show that this overhead cost is recovered if the macro-block is reused several times. However, in some cases, we can reach the objectives of time-to-market constraints from the first time of reuse.

2.2 INTRODUCTION

Macro-block reuse is the only way to build robust million-gate chips in a reasonable amount of time. For that reason, design reuse is becoming a hot topic. Several methodologies for reuse have been published, one of the first ones was introduced by Mead and Conway [Mead80] for the physical layer. Other works introduced reuse strategies and concepts that have been elaborated to facilitate the reusability of components at different levels of abstraction ([Jerr97], [Kiss97]). The main constraint to apply this technique is the fact that the reused macro-blocks are not flexible enough. To resolve this disadvantage, Preis [Prei95] proposed a library of extremely flexible parameterizable components. For the same reason, Blum [Blum93] developed a workbench for the generation of flexible quality VHDL cells and proposed a method for enabling the reuse of complex building blocks.

Some other works focused on developing efficient tools managing the reusability of designs for different purposes. For example, SISC [Giam85] is a frame-based system containing generic behavioural models of frequently used components. In [Altm94], Joachim presented a model for reusing design objects in CAD frameworks. This model considers only generic modules and multi-functional units. On the other hand, component reuse has a long-standing tradition in the use of application microprocessor based systems including RAMs, ALUs, I/Os and datapath components (e.g. adder, multiplier) [Syno94]. More recently several works report on the use of object oriented techniques for reuse ([Schu98], [Agst98]). The main emphasis is to provide a design methodology for efficient reuse. In this way Seepold [Seep96], proposed a Reuse Management System (RMS) based on the object-oriented internal data that makes easy and quick the access to the reuse data.

However, very few works concentrate on analysing the cost of reuse. The main objective of this work is to present a study aimed to analyse the cost of design for reuse on a specific application. In this chapter we discuss the design for reuse of an ATM Shaper macro block and it is organised as follows: The next section introduces the application, section three introduces our design approach and explains the difference between specific and reusable blocks, section four compares the cost of specific and reusable components, finally section five gives some conclusions.

2.3 CASE STUDY: ATM SHAPER DESIGN

This section introduces the design used in this study, an ATM shaper. ATM is a connection oriented network, where data is carried in short fixed-length packages, called cells. To provide agreed Quality of Service QoS, the network operator must be able to ensure minimal congestion conditions. Traffic shaping is one of the methods, which can be used to prevent traffic congestion in the network. This function is realized by a module called Shaper.

2.3.1 The Shaper Architecture

The Shaper device is implemented into different parts. Each of them defines a specific function that implements the traffic shaping in an ATM network system. The traffic shaping function is based on a Calendar (memory) management. Furthermore, the device can perform the traffic congestion on an Available Bit Rate (ABR) connection.

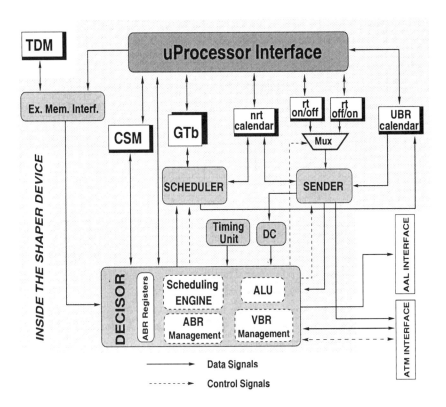

Figure 2.1 Shaper architecture

The Shaper architecture, shown in Figure 1.1, is composed of 4 interfaces, 4 main functional blocks and 7 memories containing the traffic parameters. A brief description of these elements follows. The interfaces are:
- *The external traffic description memory interface* is used by the Shaper to read the traffic parameters to comply with from the external traffic description memory (TDM).

- *The microprocessor interface* receives and exchanges control and status information with the microprocessor. The microprocessor configures the ABR parameters and it can access, in read and write modes, to all the internal memories for set up and diagnostic purposes.
- *The ATM Adaptation Layer (AAL) Interface* indicates to the shaper when a new packet has arrived and the Shaper gives to the AAL the Queue IDentifier (QID) of the cell to be sent.
- *The ATM Interface* is used for data exchange with the ATM layer and to manage properly ABR connections providing to the ATM layer the QID of the cells belonging to the ABR connections.

The internal memories used for the Shaper management are:
- Two real time (rt) calendars (RT1 and RT2) to handle the rt traffic.
- The non real time (nrt) nrt-calendar (NRT) to handle the nrt traffic.
- The UBR-calendar to handle the UBR traffic
- The Group Table (GT) memory is used to resolve the scheduling problem associated to the permit inside the NRT.
- The Current State Memory (CSM) contains the contracted timing parameters of each connection.

The only external memory used for the Shaper management is the TDM containing the traffic parameters. The main functional blocks are:
- The **Sender** reads the permits from the calendars (RT1, RT2, NRT and UBR) and sends them to the Decisor. Then the sender updates the pointers to point at the next location to be read from the calendars.
- The **Scheduler** receives from the Decisor the QID and the time stamp indicating in which memory location the permit should be stored. The Scheduler puts in the NRT calendar the permit with the help of the Group Table memory. Then, it updates the GT memory if necessary.
- The **Decisor** performs the following tasks: it validates the permits read by the Sender, and sends the right one to the proper interface (ATM or AAL), then calculates for nrt-connection the new position in the calendar of the next permit. This block also performs the ABR management in the case of ABR traffic.
- The **timing unit block** is used to control the ABR traffic parameter such as the Allowed Cell Rate (ACR) Decrease Time Factor (ADTF) and the Trm.

2.3.2 System Partitioning

We choose to partition the system into five modules as shown in the figure: the Scheduler, the Sender, the Decisor, the ABR Management blocks (ABR-reception and ABR-Emission), and the Time-unit block. Each module will be described by a single process using a VHDL behavioural description. Each process is control dominated and performs condition test and branch operations, simple arithmetic/logic operations, read and write operations and I/O operations. Since the execution of such operations is usually small, operations with dependencies can be chained into a single clock cycle by the behavioural synthesis tool. These behavioural descriptions are then synthesized using a high-level synthesis tool in order to obtain their corresponding RTL descriptions.

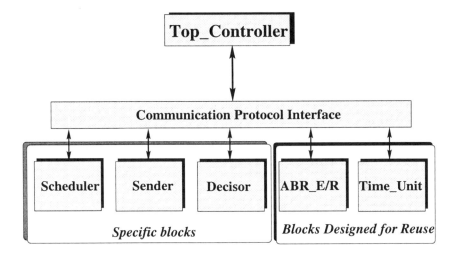

Figure 2.2 Shaper partitioning

Considering the partitioning phase, we classified the modules into two categories: To the first category belong blocks that are designed for reuse in other applications, and to the second category belong blocks that are designed to be used for the implementation of our specific application which is the Shaper device. In a behavioural description, the top controller was described, which co-ordinates these modules via a communication interface protocol.

2.4 SPECIFIC AND REUSABLE BLOCKS

This section introduces our design methodology and the difference between reusable and specific modules.

2.4.1 Modular Design Approach

One of the main methodologies applied to handle very complex designs is the structured design methodology. This was introduced by [Trim81], and is based on a hierarchical approach, which proceeds by partitioning a system into modules as shown in Figure 2.3.

Proper partitioning allows independence between the design of the different parts. The decomposition is generally guided by structuring rules aimed to hide local design decisions in such a way only the interface of each module is visible. This kind of methodology is also called "modular design", and it consists of solid design rules in terms of timing constraints, hierarchical design and floor-planning. The overall modular approach optimizes the insertion of reusable component within the circuit.

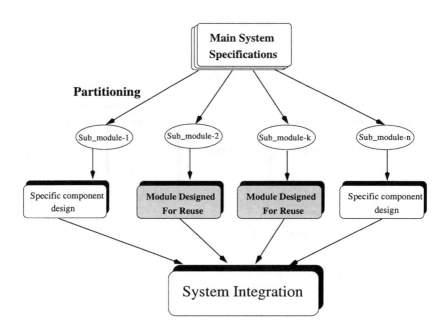

Figure 2.3 Partitioning and modular design techniques

Adapting such a design methodology, the whole system can be seen as a set of blocks designed with different methodologies and for different purposes. These purposes are the components reusability and the components stand-alone design. The number of components, which are targeted to be reused within other applications depends on the application specification. For instance, in the case of the ATM Shaper, the chip implements the ABR traffic function. Since the ATM forum has defined the standard algorithm for the ABR traffic service, it is recommended to consider the design of the ABR component as a *design for reuse* for other application. The same strategy can be taken into considerations for the Time-Base unit block.

At the same time, our application contains blocks designed for the construction of the whole Shaper, such as the decisor and the scheduler. Reusing these blocks in other applications is not planned. This is why we call them *specific components*, which are designed to be reused and only integrated in the Shaper device.

Once the blocks have been implemented and functionally verified, we developed a top-level netlist that instantiates the blocks and connects them together. This step is called the system integration step. Currently, this one is manually made, but it is recommended to develop a script that will generate the appropriate instances and automatically instantiate them in the top level netlist.

In order to achieve a high-quality insertion and integration in a new design, designers need a good knowledge of the reusable component. Thus, design for reuse concerns all additional activities that have to be performed to generate an *ease-of-use* module and a flexible module. In the next section we will present and discuss these activities for both components designed for reuse and for specific application components.

2.4.2 Design of Reusable Blocks

The key to the success of the reuse technique, is to have blocks that have been designed for reuse. Thereby, the reuse will be effective. Dealing with such a challenge, some requirements, during the design of particular blocks, have to be taken into account. This is why the effort for the *design for reuse* of a component is relatively high. Before investing lots of time in the design of a reusable component, the designer should keep in mind two particular points: 1) he or she should have good understanding about the environments in which the component will be located. 2) he or she should have an approximate idea about the number of applications in which the component will be reused. In addition to the requirements mentioned, five steps needed for the design of a component destiny to be reused in other applications:

2.4.2.1 Analyzing the application domain

From the design point of view. this point is a critical because we have to take care of all functional aspects also all the functionality expected from the module, not only in our system but when it will be implemented in other systems. In the case of the ABR block, different kind of functionalities defined by the ATM forum [ATM 96] recommendations have been considered in order to make sure that the design perfectly fits all the constraints.

2.4.2.2 VHDL Modelling and component abstraction

The basic modelling approach for reuse is the parameterization. This means that most of the required functionality needed can be controlled by parameters only.

This step takes more time for components designed for reuse than for specific components, because we have to write a *generic functionality* of the component and not a unique functionality as for some of the Shaper blocks. Doing so, we can increase the reusability and we can make a clear, concise and expressive abstraction of the reusable component.

2.4.2.3 Component Synthesis

During this step we synthesize the component according to a given budget. Each synthesizable module in the design has a timing budget. This budget is developed when the design is partitioned into sub-modules, and before coding begins. So, synthesis scripts must allow the designer to synthesize the module and meet timing goals in the final chip design. Within the modular technique this synthesis step is independent for each module. So difficult problems can be solved on small modules, where they are the most tractable. Once each module meets this timing budget, the whole design is assured of meeting its overall timing goals. This step is almost the same for specific and reusable components.

2.4.2.4 Component verification and test-bench generation

This step is performed by making an intensive simulation of the functionality. In the case of reusable components this step has to check all the possible scenarios where the component can be used. As it will be shown later, this may induce a lot of overhead time when compared to the verification needed for specific components. The designer performs this verification step using a test-bench, which is typically based on a functional model which allows to generate test suites in a flexible way.

2.4.2.5 Documentation

We need to create a complete set of user documentation to guide the designer in using these modules to develop the chip design.

In the next section we will show that steps 1, 4 and 5 are much cheaper for the specific components when compared to reusable components. Thus the cost of design for reuse is higher than the cost of specific component design.

2.5 COMPARING REUSABLE AND SPECIFIC COMPONENTS

As it is stated above, the modules designed for the ATM Shaper are splitted into two categories. The first one includes blocks that are going to be reused for other applications. These are the ABR block and the Time-Base-Unit block. These blocks can be considered as modules *designed for reuse*. The second category concerns blocks that are designed to be used for the construction of our ATM Shaper. The scheduler block and the sender block belongs to the second category. These blocks, as we will see later, take less time for design than the blocks designed for reuse.

Category of Blocks	Blocks "Designed For Reuse" in other Systemes		Blocks designed to be reused for the construction of the whole Shaper	
Blocks	ABR_Block	Base_Unit_Block	Scheduler_Block	Sender_Block
Analysing the Application domain	40	20	8	8
VHDL Modeling Component Abstraction	30	25	25	30
Component Synthesis	30	20	25	25
Verification and TestBench generation	70	45	35	30
Documentation	80	30	7	7
Total design Cost	250%	120%	100%	100%

Table 2.1 Cost of design

Table 2.1 shows, in terms of time, the design evaluation cost for some of the Shaper modules. This table gives the fraction of the effort needed for each step. In order to make the comparison easier, we normalized the costs. All the cost are related to the standardize time required by an experienced designer to develop from scratch a single instance of a component. This time is set to 100%.

The results shown in Table 2.2 demonstrate that the time spent for developing a *design for reuse* component may be much higher than the time spent for designing a stand-alone component. In the case of the ABR module, the design for reuse is 2.5 times larger than the design of the same block for a specific application. Of course this ratio may be lower for simple blocks like the time-base unit module because of the lower complexity of this module. The main steps that extensively increase this overhead of time are the application analyzing domain, the component verification with test-bench generation and the documentation. The latter has not the same signification for a component designed for reuse as for a component designed for a specific application. In the first case documentation means *user guide manual*, while in the sccond case it means *comments* written in the VHDL description file describing the functionality of the module. Concerning steps 2 and 3, we can see that the time spent for the VHDL modelling and the component synthesis is not much higher for a reusable component than for a specific component.

Blocks	Blocks Designed for Reuse		Specific application blocks	
	ABR_Shaper To ATM	Base_Unit	Scheduler	Sender
Gate count	4300	800	5400	4800
Men months	1,8 M/month	1/4 M/month	1 M/month	1 M/month
Productivity gate/Month	2388	3200	5400	4800

Table 2.2 Design productivity

Table 2.2 shows the complexity and the design time of the modules used. This table shows also the productivity for reusable and specific components. Of course the productivity depends on the complexity of the module. This table shows that for modules of similar complexity, productivity is twice as high for blocks designed for a specific application. One should note that in all cases, productivity is high, mainly because the design was started by a behavioural synthesis and using high-level synthesis.

In order to show the real cost of reuse, we will analyze the case of the ABR block. In this analysis we assume that the reuse of such a block in another application will take 20% of the time required to design the ABR for specific application. This lapse of time includes [Krue92] the elaboration and the adaptation of the block.

Figure 2.4 shows the cost of N redesign versus one design and N reuse. In this case, it is clear that the ABR needs to be used more than 3 times in order to justify design for reuse. Of course the numbers are very specific to this block. For example, the same hypothesis on the cost of reuse leads to a better pay off of reuse in the case of the Base-Unit block. In fact, for that case, reuse is profitable even if the block will be reused only once.

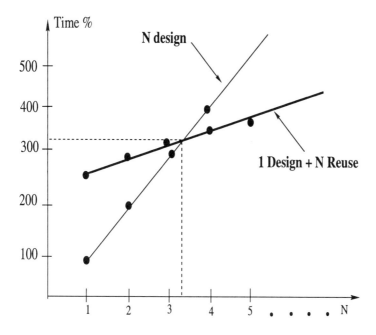

Figure 2.4 Cost of design versus reuse

2.6 CONCLUSION

In this paper, we have demonstrated that design reuse can effectively enhance design productivity and improve system time-to-market constraints. This work was based on some measurements made during the design of a chip containing 2 million transistor, namely an ATM Shaper. We have shown that modules designed for reuse may be up to 2.5 times consuming more time than modules designed for a specific application. This is mainly due: First, to the analysis and considerations made for the modules to be reused in other applications; Second, to the extra verification time and to the test-bench that increases in complexity to ensure total functionality and validation; and third of course, to the documentation.

Furthermore, we have shown that the overhead of time spent in the design of a reusable component is recovered if we reuse the component several times. This number depends on the complexity of the component. In some cases we can reach the objectives of reuse at the first reuse time. Meanwhile, it is very important, before making design for reuse, to consider the number of applications in which the block will be reused versus the overhead of time to get a reuse macro block.

Imed Moussa received his Engineering degree (1991) from the National School of Engineering, Tunisia, his M.S, degree from the University Blaise Pascal, France, and his Ph.D. degree (1996) from the National Polytechnical Institute of Grenoble (INPG), France. Since 1996 he has been a member of the System Level Synthesis group and research engineer at TIMA Laboratory where he has developed and designed VLSI telecommunication circuits for ATM applications. He has participated in the program commitee of several conferences and a workshop. His research interests include hardware engineering, large scale communication system design and IP reuse.

Ahmed Amine Jerraya received his Engineering degree from the University of Tunis at Tunisia in 1980, a Docteurs Ingenieur degree in 1983 and a Docteur d'Etat degree in 1989 from the University of Grenoble in France. He has held a full research position with the CNRS in France since 1986. He has participated in several successful French CAD projects such as the LUCIE project in the early 80's and more recently the high-level synthesis system AMICAL. He has served as a program and general chair for the 8th and 9th International Symposium on System Synthesis (1995,1996). He has also served as Co-Program and Co-General chair of Codes workshop (1998, 1999). He is the vice general chair of DATE 2000. He is currently the head of the system-level synthesis group within TIMA Laboratory where he is developing methods and tools for behavioral synthesis and hardware-software codesign.

Mario Diaz Nava received his B.S. degree in Communications and Electronics Engineering from the Instituto Politecnico National, Mexico, in 1978, his M.S. degree and Ph.D. degree in 1982 and 1986, both in Computer Science, from Institut National Polytechnique de Grenoble (INPG), France. Since 1990, he has been with STMicrolectronics where he has analysed and designed communications architectures and integrated circuits. From 1986 to 1990 he worked for APTOR as senior engineer. He was the project leader on such circuits as the ATM-SDH Line Termination and AALs, and was contributing to several European projects RACE (Technology for ATD, COMBINE), Esprit (CANDI, TIBIA), and JESSI (AE97, AE102, AE103). He is currently Hardware Design Manager at Multimedia Networks Strategic for VDSL and Network Termination Advanced Telecommunications I.C.s.

3 A FLEXIBLE CLASSIFICATION MODEL FOR REUSE OF VIRTUAL COMPONENTS

N. Faulhaber and R. Seepold

FZI Karlsruhe
Department Microelectronics System Design
Karlsruhe, Germany

3.1 INTRODUCTION

The design of microelectronic systems is heavily driven by the fact that transistor and feature size have constantly decreased, while frequency and density have increased. This gain has been supported by the development of new technologies and manufacturing equipment, which provide mechanisms to improve design efficiency.

Since the application of a design reuse methodology is expected as the key enabler to manage complex designs to an aggressive time-scale, efficient classification and retrieval mechanisms have to be developed to support design reuse of Intellectual Property (IP). Therefore, the core of a reuse management system has to support both design for reuse and reuse of design capabilities.

Customizing IP according to its functionality is an inherent objective of these design capabilities. In order to meet these objectives, a virtual component (VC) that contains IP must be classified to enable proper storage in a reuse database, and furthermore, a classification-based retrieval has to be implemented that ensures access and customizing of a VC to the given specification.

3.2 OBJECTIVE

In order to consequently support reuse, available IP has to be classified and stored for reuse. An essential question is how IP reuse is supported by a reuse management system's functionality. The answer to this question is closely related to the methodology that is used to store and retrieve reusable components.

Many existing systems concentrate on very specific classification techniques and retrieval algorithms dedicated to allocate a desired component or functionality [Jha 95]. In parallel, customized descriptions of meta-data are provided by structures that are restricted to the underlying core model.

An analysis has shown that four major classification and retrieval mechanisms are applied: taxonomy, key words, attributes and similarity metrics. Although each mechanism has certain advantages, no system or methodology is documented that tries to combine the strength of each approach, and therefore, to summarize all mechanisms available for IP reuse in one model.

Integration support of existing VCs into a reuse database, support of design for reuse, support of aggregation of VCs to new and more complex VCs and support of reuse of VCs stored in a distributed database environment are some of the main advantages of the available reuse system RMS that will be extended [Seep98].

The main objective of the work presented is the development of an efficient classification model to manage different classes of IP and to support appropriate retrieval mechanisms to get access to exactly fitting and similar IP. A concept will be developed based on a classification model to specify attributes and similarities, while maintaining queries, i.e., based on key words, and a powerful taxonomy. This extends the existing reuse management system (RMS). The aims of the whole work can be summarized as follows:

- close the gap of existing reuse systems and classification techniques
- develop a flexible classification model for VCs
- provide efficient retrieval mechanisms
- implement the concept on top of a commercial database

3.3 STATE OF THE ART

The development of an extension for RMS has been started after the results of the state of the art analysis have been available. In the main focus of the study was the detection of the mechanisms applied, and the evaluation of advantages and disadvantages for each mechanism. Therefore, 12 models have been selected for evaluation ([Agst98], [Chou96], [Desi98], [Klei98], [Lebo97], [Mart97], [Oehl98], [Olco98], [Seep98], [Sehg94], [Sieg97], [Wage95]).

Each model can be regarded as a typical representative of a class of reuse models. As a result, four different mechanisms could be isolated that are applied to reuse component storage and retrieval: taxonomy, key words, attributes and similarities. While some of the models do not even support flexible IP, none of the models support more than two mechanisms. The detailed evaluation can be found in [Faul98].

The advantages and disadvantages of the different mechanisms are summarized in Figure 3.1. Obviously, none of the mechanisms can offer all advantages while avoiding major restrictions. For example, the clear structure that is provided by a taxonomy is not available for similarity based mechanisms.

	Taxonomy	Keywords	Attributes	Similarities
Advantage				
• Clear structure	●			
• Intuitive search	●	●		
• Simple query specification	●		●	○
• Convenient browsing	●			
• Quick classification		●		
• Flexible mechanism			●	●
• Extendable structure	●	●	●	●
• Ranked component list				●
Disadvantage				
• Covers only one feature				
• Important features missing	●	●		
• No similarities	●	●	○	
• Ambiguity in classification	●			
• Missing taxonomy/structure		●	●	●
• No browsing possibility		●	●	●

Figure 3.1 Evaluation of retrieval mechanisms

As an extension of this study, a criteria catalogue has been defined describing key features to be supported by the new approach. In detail, the requirements are stated in the following list:
- provide a flexible and extendable classification scheme
- support structured classification and browsing (known from a taxonomy)
- capture all features of an IP (like attributes)

- enable definition of similarities
- provide quick integration of new IPs
- enable intuitive query specification and search
- deliver a ranked list of matching and similar IPs

In the following sections, a new classification methodology is presented that will extend the Reuse Management System (RMS) that has been developed at FZI. This extension will be based on a flexible and customizable taxonomy that is accompanied by attributes and a powerful similarity metric.

3.4 RMS SIMILARITY METRIC

The similarity metric developed for the extended RMS-Classification is based on the methodology applied in the Reuse Management System (RMS) [Seep98]. Due to the fact that RMS has a very flexible object-oriented data model for managing IPs, RMS has been extended with a new classification methodology that offers both the combination of all four mechanisms and the fulfilment of the key features of the criteria catalogue. The model extension will be called *RMS-Classification*. RMS-Classification offers two new concepts: asymmetric similarity and conceptual similarity. In the following, the basic similarity metric and the terminology are introduced, while both new extensions are discussed in detail.

3.4.1 Basic Similarity Metric

In the RMS terminology, a reusable VC is called a Component Environment (CE) and a Characteristic Attribute (CA) represents a feature like *add*, *multiply* etc. The association of a CE to a CA is provided by a weight, which indicates how well the CE fulfils the feature of the CA. Furthermore, CAs can be considered as classes of reusable IPs (cf Figure 3.2).

Subsets of CAs with a certain similarity relationship between them are joined by Vectors of Characteristic Attributes (VCA). The level of similarity of a CA in a certain subset is represented by its weight in the corresponding VCA. The similarity between two CEs is given by their *conceptual distance*, which is calculated as the sum of the four weights on the path between the CEs. For example, the similarity between CE_1 and CE_2, concerning VCA_1, is calculated as the sum 10+5+5+5=25 (cf Figure 3.2). VCAs can be defined for different subsets of CAs. If two CEs are similar concerning more than one VCA, the maximum of the accumulated weights will be taken for computation of the similarity. The similarity between CE_1 and CE_2 in Figure 3.2 is the maximum of the two similarities concerning VCA_1 and VCA_2.

All VCAs constitute a Characteristic Vector (CV). The CAs and VCAs within a CV describe the components that are associated to one feature, for example the feature *Function*. Within a CV, each component is exactly assigned to one CA.

A complete classification scheme can consist of several CVs (cf Figure 3.2). This allows to describe reusable components by defining a similarity related to different features. Each reusable component is exactly assigned to one CA per CV. For more than one CV, the similarity between two components (CE) is described as a vector that consists of several similarities for each CV. For example in Figure 3.2, the similarity vector for CE_1 and CE_2 is given by the vector (40, 55).

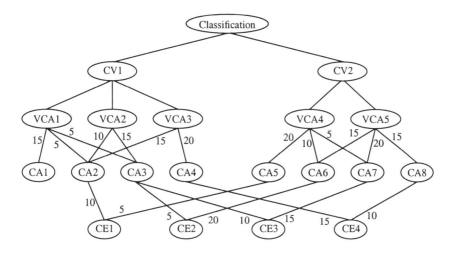

Figure 3.2 RMS similarity metric

3.4.2 Conceptual Similarity

In the basic RMS metric, similarity is only defined between CEs, i.e., this is the conceptual distance. The first extension of this concept is called *conceptual similarity* and it defines the similarity between two CAs related to the same VCA as the sum of the two available weights. If two CAs are similar concerning different VCAs, the path with the maximum of the weights is considered. Thus the similarity between CAs is implied by the similarities of CEs. This does not define any new connection between CAs and VCAs. Since CAs can be viewed as classes of components, this concept allows the definition of similarities between classes of VCs. Furthermore, this concept is used to define similarities between internal nodes of a taxonomy.

Figure 3.3 shows a set of different CAs that are connected through VCAs. The conceptual similarity between CA_1 and CA_2 is 25 that is the sum of the weights on the edges via VCA_1. The similarity between CA_2 and CA_3 is given by the maximum of 35 and 15.

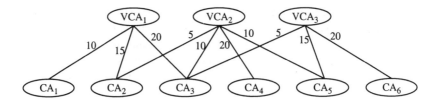

Figure 3.3 Conceptual similarities

3.4.3 Asymmetric Similarity

In the basic RMS similarity metric, the similarity between reusable components is defined as a symmetric relationship. But several asymmetric relationships can occur: For example, an ALU can be used instead of a multiplier but not vice versa. Exactly the same is true for an adder and an ALU or an adder and a multiplier, because a multiplication might be realized by repeated additions. Initially, the status can be described as stated in Figure 3.4.

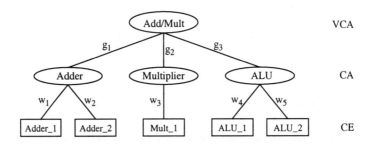

Figure 3.4 Example of asymmetric similarities

If an asymmetric similarity is considered between CAs of a VCA, the final similarity relationship corresponds to a partial order. A partial order is equivalent to a directed acyclic graph (DAG). Therefore, each VCA is extended by a DAG description that explains asymmetric similarities between related CAs. Those CAs, which are not included in this DAG, are symmetric similar to each other CA within its VCA.

Figure 3.5 introduces the DAG of the VCA of Figure 3.4 and it introduces the symbol used in RMS. The edges of this similarity DAG imply that an ALU can be used instead of a Multiplier or an Adder and an Adder can be used instead of a Multiplier. Since a partial order is transitive, the edge between Multiplier and ALU can be removed without loss of information.

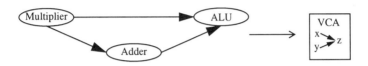

Figure 3.5 DAG and symbol of a VCA

3.5 THE RMS-TAXONOMY

Initially, the core of the RMS-Taxonomy was similar to the „Yellow Pages" [Desi98]. But two important differences have to be mentioned: The RMS-Taxonomy is strictly hierarchical, and therefore, it can be represented by a tree. The second difference is the clear distinction between taxonomy and attributes. For example, whether a DSP is an 8, 16 or 32bit DSP can be better covered by a single attribute instead of three sub trees. A further example is the distinction between components which work on different data types. Therefore, a better solution is presented in Section 3.6.2

3.5.1 Taxonomy Structure

Through successive levels, the RMS-Taxonomy refines the classification of the VCs according to their functionality. At the top level, VCs are subdivided into disjunct classes of components, e.g., processors, controllers, buses etc. Each of these classes can be refined into several subclasses. The relationship between an upper and a lower class is provided by an *is-a* semantics [Rumb94]. Therefore, a lower class simultaneously is of the upper class type. This refinement can be recursively performed and it spans a tree that can be viewed as a specialization hierarchy. Figure 3.6 shows a part of the RMS-Taxonomy. The classes of the taxonomy are represented by ellipses.

A VC is classified within the RMS-Taxonomy by assigning it to the leaf node which best characterizes the VC's functionality. In Figure 3.6, the classified VCs are represented by rectangles. Due to the *is-a* semantic between a taxonomy node and his parent, each node can be considered as the generalization of its children nodes. Furthermore, each internal node gets assigned the union of all IPs that are

28   REUSE TECHNIQUES FOR VLSI DESIGN

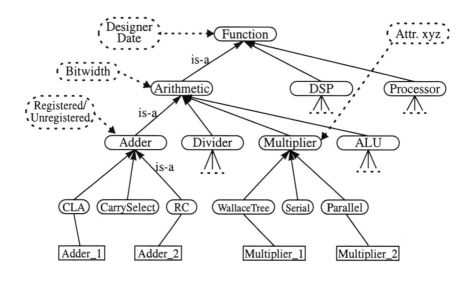

Figure 3.6 RMS-Taxonomy

associated to its children nodes. The class Adder in Figure 3.6 contains two VCs as CEs: Adder_1 and Adder_2.

3.5.2 Attribute Assignment

Besides all advantages, taxonomies in general have one disadvantage: Since they only cover one feature of the classified components (e.g., the function), other features, like designer information cannot be covered. In the extended RMS-Classification, those features are described by attributes. However, an adder is described by other attributes than a DSP because it is not possible to define a common and comprehensive set of attributes.

Therefore, in the extended RMS-Classification, the following procedure has been realized: Attributes are assigned to a taxonomy node if these attributes are relevant for all children nodes and if these attributes are not already assigned to a parent node. For example, attributes like designer or project name are relevant for all VCs. They are assigned to the taxonomy root. Other attributes that are only relevant for adders, are assigned to the node ADDER.

A VC, classified by connecting it to a leaf node, is characterized by the values for all the attributes assigned to the nodes on the path from the taxonomy root to the corresponding leaf node. This allows definition of specific attribute sets for different classes of VCs. For the taxonomy in Figure 3.6, attributes are defined for different taxonomy nodes. These attributes are represented with dashed lines (e.g., Designer). Using this example, for a CarrySelect adder the following attributes can be set: Designer, Date, Bitwidth and Registered/Unregistered. For a parallel multiplier, the attribute set consists of the attributes assigned to the nodes Function, Arithmetic, Multiplier and Parallel.

3.6 EXTENDED RMS-CLASSIFICATION

With the help of the extensions introduced, VCs can be classified by a taxonomy and relevant attributes can be specified to increase the granularity but a similarity metric is still missing. In the following, the RMS-Taxonomy is coupled with the previous presented similarity metric.

3.6.1 Taxonomy and Similarity

The leaf nodes of the RMS-Taxonomy can be regarded as classes of IP. In the similarity metric of RMS, CAs had been introduced as classes of IP. To relate CAs to the taxonomy leaves, the assignment of an IP to a taxonomy leaf is done by means of a weight to indicate how well it fulfils the respective functionality.

Now, taxonomy leaves can be considered as the CAs of a CV, e.g., this CV describes the function. For the CAs associated, VCAs can be defined as presented in the RMS similarity metric. Due to that concept, similarities between IPs can be calculated. For example, similarities between CLA, RC and CARRYSELECT can be defined by connecting them to a VCA and introducing a DAG if they are asymmetric similar (cf Figure 3.7).

Due to the *is-a* semantic between taxonomy nodes, internal nodes can be considered as classes of components, too. Therefore, internal nodes are identified as CAs and they implicitly define conceptual similarities between them. It is possible to include in a VCA both leaves and internal taxonomy nodes. However, it is not possible to define similarities between two taxonomy nodes which are a direct or a transitive member of a parent-child relationship. For example, conceptual similarities can be defined between ADDER, MULTIPLIER and ALU by adding a new VCA with its own DAG.

3.6.2 Additional CVs

Besides the CV *Function*, this similarity metric offers the possibility to define further CVs, like *Input* or *Output*. The administrator decides whether a feature should be integrated as an attribute or a node of the taxonomy. The definition of a taxonomy node has the advantage that similarities can be defined on that structure, and furthermore, an adoption of the VC can be performed via filters. The modelling in terms of an attribute has the advantage that the taxonomy is not changed. Via the user interface, attributes can be accessed via specific windows.

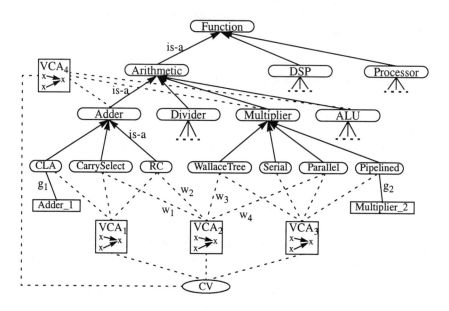

Figure 3.7 Extended RMS-Classification

3.6.3 Retrieval of Intellectual Property

For the preparation of a query, the user specifies the function of the IP required by browsing within the RMS-Taxonomy or by selection from a key word list. In a second step, attributes might be specified that are available for the selected taxonomy node. In a third step, the search procedure can be initiated.

As presented in Figure 3.8, the system searches for IPs that are matching the specification. In detail, each IP is assigned to a taxonomy node. The matching IPs are assigned to the node selected. The algorithm searches for similar components. That means that starting from the selected node all children nodes are evaluated according to the similarity metric. If a similar CA is a taxonomy leaf, the IPs assigned are considered as similar IPs. If a similar CA is an internal node, CA is considered as conceptual similar. For matching and similar IPs, the attributes are checked in a third step. If a specified attribute is not fulfilled by a retrieved IP, the IP is removed from the final list. For similar IPs, only the relevant attributes are checked. In the final step, the result is presented to the user. Matching and similar IPs are ordered corresponding to their similarities.

```
Search_VC (CA_F, CA_I, CA_O) {    // Function, Input, Output
    Erg_VC := ∅; // initialize
    Erg_CA := ∅;
    if (CA_F == taxonomyLeaf) {
        Erg_IP := getMatchingVCs(CA_F, CA_I, CA_O);
        Erg_IP := getSimilarVCs(CA_F, CA_I, CA_O) U Erg_IP;
        Erg_IP := checkAttributes(Erg_IP);
        Erg_CA := getConceptualSimilarCAs(CA_F);
    } elseif (CA_F == internalTaxonomyNode) {
        for (allLeafs CA_j of CA_F) {
            Erg_IP := getMatchingVCs(CA_j,CA_I,CA_O) U Erg_IP;
            Erg_IP := checkAttributes(Erg_IP);
        }
        for (allNodes CA_j of CA_F) {
            Erg_IP := getSimilarVCs(CA_j, CA_I, CA_O) U Erg_IP;
            Erg_IP := checkAttributes(Erg_IP);
            Erg_CA := getConceptualSimilarCAs(CA_j)U Erg_IP;
        }
    } endif
    if (Erg_IP == ∅ AND Erg_CA == ∅) {
        Search_VC (Parent(CA_F), CA_I, CA_O)
    } endif
}
```

Figure 3.8 Retrieval algorithm

3.7 IMPLEMENTATION

The implementation has been performed on the commercial database (Oracle™). The development of the application has been performed in two major steps. The first step has been the definition of the database using the *Designer/2000* CASE environment: Designer/2000 offers a comprehensive development environment which consists of a set of tools that support all relevant design steps. Some of these steps have to be manually performed, others can be automatically done. Starting with a requirement analysis, each database view (conceptual, logical and physical) has been developed and implemented until the final database could be compiled. The transfer of RMS classes and taxonomy classes has been manually performed. The details of this process are beyond the scope of this article [Faul98].

In the second step, the development of the application with *Developer/2000* tools is presented: For the development of the application, the *Forms Designer* has been applied. The according *forms-application* can be executed as a client/server application that is running in an Intranet or Internet environment. For both alternatives, Oracle™ provides a specific architecture that consists of the desktop machine, an application server and a database server as the backbone. The user interface is build up during run-time so that each Java-based Internet browser is able to compute the output, and therefore, the GUI is platform independent. Each action within an applet is transferred to the application server that computes the interactive input of the user.

The principle architecture is stated in Figure 3.9.

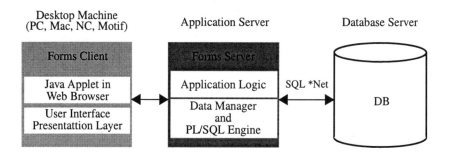

Figure 3.9   ORACLE web architecture

3.8 APPLICATION OF THE MODEL

In order to demonstrate the efficiency of the model developed, a design of a processor has been performed. Therefore, several alternative implementations of virtual components have been integrated into the RMS database. Each design integrated can be selected during the design process, and this preparation simulates the status of the database that might have been achieved during a regular design for reuse application.

Besides the integration of virtual components, a classification scheme has been developed. The classification scheme is based on the model that has been presented in the previous sections. For this application example, the classification scheme consists of the CVs *Function*, *Input* and *Output*. The nodes of the taxonomy have been extended by attributes which are relevant for this example. Initially, the weights of the relationships between CAs and VCAs have been fixed to 100, but they could be re-defined during the application of the system, in order to represent a higher precision degree (cf. Table 3.1).

The full flexibility of this approach can be used while defining the input and output characteristics by CVs instead of coding them by attributes. This fact prevents the designer from introducing redundant sub trees in a complex taxonomy. The CVs *Input* and *Output* have been separately modelled (cf. Table 3.2).

A FLEXIBLE CLASSIFICATION MODEL

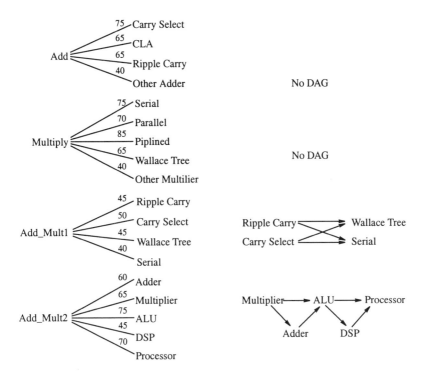

Table 3.1 VCAs of the CV *Function*

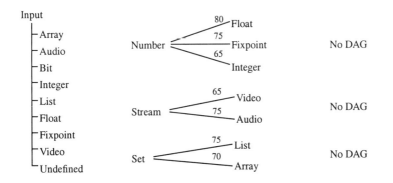

Table 3.2 CAs and VCAs for *input/output*

For example, the processor should contain an ALU that should not be developed from scratch. Therefore, the designer specifies the ALU with the main specification window (cf. Figure 3.10).

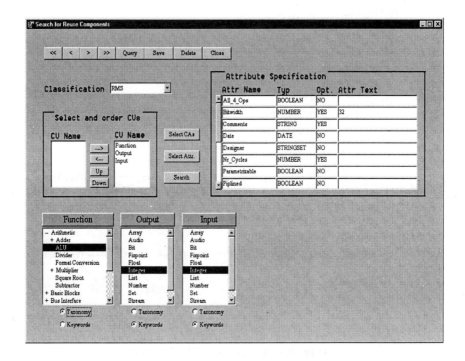

Figure 3.10 Specification of the ALU

Inside the taxonomy function, the component type ALU is selected. The input and output types are integer and the bit width is set to 32. After starting the search algorithm (search button) for this specification, three candidates will be offered for reuse: ALU_1 (fitting component) and ALU_2 and ALU_3, which are similar components. In this case, the designer might decide to include the fitting component into his design. In a final step, he might check the additional (not specified) attributes of the component and based on the results of this check, it might be possible to decide to reject this component and to inspect the similar components. For each component computed by the search algorithm, a separate window is available to view the attribute values (cf. Figure 3.11).

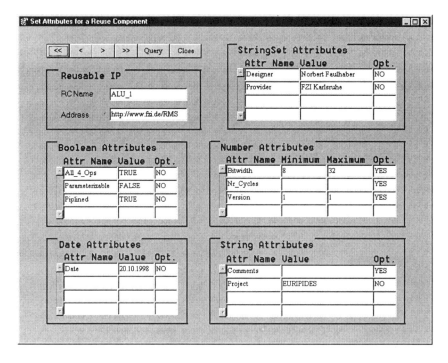

Figure 3.11 Attributes of ALU_1

3.9 CONCLUSION AND OUTLOOK

The extended RMS-Classification model closes the gap of existing models by providing a methodology that supports four classification mechanisms: taxonomy, key words, attributes and similarity metrics. In parallel, the flexibility of the basic RMS-similarity metric could be maintained, and therefore, it is possible to customize this system to specific requirements mainly defined by the IP that should be made reusable.

Due to the fact that the same information can be accessed by different methodologies, the classification scheme can be customized, too. According to the needs, the administrator, which is the main person defining the classification scheme, may decide to introduce a characteristic by attributes or by a taxonomy node. In case of modelling the characteristic by taxonomy features, each node could have different attributes and similarities.

The similarity metric presented offers very flexible relationships to define similarities between components: Besides the application of common similarity computation, the term conceptual similarity has been introduced to describe the possibility of managing similarities at a higher level of abstraction within the underlying taxonomy. In order to maintain the classification mechanisms for component retrieval, all four features have been integrated into the search algorithm. The search process based on taxonomy or key words can be selected by the designer without any loss of information. In parallel, the specification of additional attributes may be selected where required.

The implementation of the system has been performed for an Oracle™ database that supports the access from Java-based clients which might can be accessed via the Intranet or Internet.

In the future, the extension of the classification by a thesaurus tool might be a separate task. Further investigation could be performed in enhancing the similarity computing by introducing attributes into the similarity calculation.

Norbert Faulhaber received his Diploma degree in Computer Science from the Technical University of Karlsruhe in 1998. His research interests include database systems, reuse techniques for microelectronic system design and Internet/Intranet applications. Since December 1998, he works at the Internext Consulting Group, Karlsruhe, Germany.

Ralf Seepold received his M.S. degree in Computer Science from the University of Paderborn in 1992 and his Ph.D. degree from the University of Tübingen in 1997. In 1992 he joined the computer science and research center FZI in Karlsruhe. In 1997, he became project leader, and since 1999, he is head of the department "Microelectronics System Design". In 1998, he became director of the German ECSI Office at FZI that is a representative of VSI Alliance. He served as a founder and general chair of the "Workshop Reuse Techniques for VLSI Design" in 1997 and 1998, respectively. Since 1998, he is member of the steering group for the "MEDEA/Esprit Conference". In 1998, he also served as a topic co-chair for "Design Automation and Test in Europe" (DATE) on design reuse, and in 1999, he is topic chair for "Design reuse and IP" at DATE, organizer and chair of the VSIA session on standardization at DATE, program chair of the "Forum on Design Languages" (FDL) and general chair of the "Workshop on Virtual Components Design & Reuse" (VCDR). His research activities include methodologies for design reuse management systems and related fields.

4 AN INTEGRATED APPROACH TOWARDS A CORPORATE DESIGN REUSE STRATEGY

U. Schlichtmann and B. Wurth

Siemens AG,
Munich, Germany

4.1 INTRODUCTION

Semiconductor manufacturing process technology continues to progress relentlessly. The 1997 release of the SIA roadmap predicts 0.15µm process technology, with the potential to integrate 40 million transistors on logic chips, for the year 2001 [Asso97]. The acceleration of the introduction of process technologies is evident from the fact that the 1994 release of the SIA roadmap predicted only 0.18 m process technology for 2001 [Asso94]. As was recently pointed out [Nepp98] that the semiconductor industry may reach the limits of silicon technology. Today, however, this date appears to be well into the future. Structures that were thought to be impossible using traditional lithography are being routinely fabricated in 0.25 m fabs already. The introduction of 300mm wafers is well on its way.

For the near future, the process technology roadmap appears to be achievable, even faster than originally thought. This puts other challenges for the semiconductor industry into the spotlight - challenges that were not expected a few years ago. The major issue today is productivity of designers - determined by design methodologies and the capabilities of EDA tools. With process technology capability (measured in transistors per chip) increasing at a rate of 58% annually, but design productivity increasing at 21% annually only, the infamous "design productivity gap" results. Design teams would multiply in size unless productivity dramatically increases, which again reduces the productivity of the individual design because of increased communication overhead. The applications for multi-million-gate chips are definitely available - markets such as highly parallel, high-performance processors; graphics engines; mobile communications chips with speech recognition capabilities all call for high-complexity ICs. And beyond such applications, entirely new markets will be opened up by chips emulating capabilities reserved for humans, such as pattern recognition. Recent publications ([Keat98a], [Payn98]) point out scenarios of System-On-Chip (SoC) chips comprising dozens of cores, up to even 100 cores on one chip.

The solution, unless we want to restrict ourselves to chips containing just about only memory, a processor core and very little other logic, is to move to higher levels of abstraction. System chips of the (near) future need to be designed from predesigned functional building blocks. These building blocks are referred to as IP (Intellectual Property) cores or simply cores. The vast majority of these cores will need to be independently designed from a specific application chip. They will then be used in a number of application chips. The teams designing the cores will be separate from the teams integrating cores into a chip, today this is the case for standard cell, memory and I/O libraries. Also, no company will be able to create and maintain all the cores required for its designs. Business models need to be developed to enable exchange of cells containing proprietary knowledge between companies - even between different business units within one company.

Within the scope of its "Logic Initiative", Siemens Semiconductor started a major effort to significantly increase design reuse. This effort is known as project "Core-Based Design". The goal of Siemens Semiconductor was to create a coherent strategy, rather than just to develop individual components of a reuse concept. We are convinced that only by turning such a coherent overall strategy into reality we can reap the promised benefits of design reuse.

In this paper, we will focus on the non-technical issues of turning design reuse into reality. For some background, we will briefly discuss our view of the difficulties of design reuse. Then we will introduce the Siemens Semiconductor process for creating reusable IP. A corporate organization to support this process is important. An intra-company business model presents formidable challenges. Creating and assuring quality is an especially important cornerstone of a corporate reuse strategy. IC design flows require significant enhancements for reuse to work. Finally, we will show how we pull all these ingredients together for the Siemens Semiconductor corporate reuse strategy.

4.2 WHY IS DESIGN REUSE DIFFICULT?

Contrary to popular current sentiment, design reuse is not a new concept. When the semiconductor industry moved from transistor-level design to cell-based design and designers began to integrate predesigned memory blocks onto ICs, this already constituted a very successful form of design reuse. This is still the standard approach today.

But now the industry needs to rapidly move and on a broad base to reusing significantly more complex logic functionality: instruction set processor (µC, DSP) cores; peripherals, graphics engines; standard interfaces such as PCI, USB and FireWire; encoding and decoding blocks; all kinds of filters etc. Analog blocks as well, e.g. ADCs, DACs, OpAmps, PLLs etc. This represents a significantly more challenging approach to reuse than standard cells and memories. One reason is that the functional variety is much broader. Therefore, it is not easily possible to form a separate organizational entity that defines, develops and supports functional cores. Reuse needs to be a driving goal throughout the entire company. Therefore, a change in culture is required. Another reason is that the move to design reuse needs to be accomplished in very short time.

The challenges here appear on multiple levels:

Core Portfolio Concepts: There is not much use in just having a large library of cores. The library needs to exactly contain those cores that are essential to the application markets that a company is serving. Significant effort is involved in making a given functionality available as a reusable core. So this won't be done unless there is a clear business case for reusing the functionality under consideration. Also, there is not much use in having a large library of cores for the right applications, if they do not fit together. For cores to fit together well, e.g. bus interfaces, clocking strategies, power strategies, interrupt concepts, OCDS, reset concepts all need to be standardized.

Core Functionality: Typically, today's blocks are developed for one chip project only. There is no incentive for the designer to consider other projects' requirements. A block developed in such a way needs to be adapted and generalized for reuse (e.g. different interfaces, clocking).

Core Quality: What exactly are the core deliverables (CAD views, documentation) that enable easy reuse? Who provides these deliverables?

Core Availability: How does a designer find the company's cores with all their deliverables? Is there an easy-to-use mechanism to deliver the core to the integrator?

Core Support and Maintenance: Who provides 1st level and 2nd level support during core integration? Who incorporates core bug fixes and updates (new deliverables, new technologies)?

4.3 CORE SUPPLY PROCESS

A cornerstone of the Siemens Semiconductor reuse strategy is the **Core Supply Process**. For a long time already, IC products at Siemens Semiconductor have been developed according to *a Product Definition and Development Process (PDDP)*. The PDDP streamlines product definition and development. This is achieved by defining checkpoints for key decisions and handoffs, by clearly assigning responsibilities for specific tasks, and by introducing working groups that handle the complete process for a product. Due to the PDDP, decision-making is more transparent, and time and effort for specific definition and development phases can be measured. This allows to identify process bottlenecks and to improve overall project planning. The PDDP helps to increase productivity, to shorten the development cycle, and to improve product value.

Now, in the emerging SoC era, designers will increasingly specialize: they either perform core creation or chip integration. Since core creation work significantly differs from chip integration work, we introduced a new process. While chip integrator's work is covered by our traditional PDDP, the new *Core Supply Process* captures all tasks from the inception of a core idea through core creation to core support. A key distinction of the CSP with respect to the PDDP is that the market for a core is company-internal, and therefore, requirements for a core come from many business units. The CSP covers core development starting either from scratch or from existing blocks (*core productization*). While the process also covers the latter case, it is our experience that it requires significant effort to generalize the functionality of an existing block that was not designed for reuse. This effort in some cases even exceeds the effort for development from scratch. Therefore, we see the process for core development from scratch as the primary process variant. Core productization, however, can be successful if it only requires the development of missing deliverables such as CAD views and possibly minor changes to functionality.

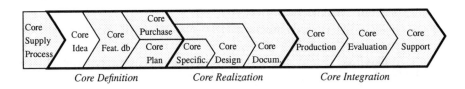

Figure 4.1 Core supply process

The Core Supply Process (cf Figure 4.1) consists of the three top-level phases shown here, each of which comprise some more detailed steps. The first three steps relate to the definition of a core, followed by the actual design of the core, which leads to the integration of the core into chip design projects.

Obviously, it is not sufficient to just develop such a process and then have its use mandated by management. Doing this just creates some "push" for designers, definition people and the other concerned parties. Some "pull" is required as well. This we achieve by providing training on the new process, distributing CSP handbooks with useful (and mandatory!) checklists, and actively supporting the application of the CSP. The CSP handbook clearly spells out the responsibilities of the various organizational units in each phase of the process.

Core Definition Phase: This process phase is crucial for successful design reuse. If the functionality of a core is not useful in a certain context, there will be no design reuse. Therefore, requirements of potential internal core customers (i.e. design projects within the business units) must be known early to take them into consideration during the core definition. Core definition people must therefore communicate with product definition teams in several business units, but the coordination effort must go even further. To achieve maximal reuse of a core in the future, requirements of future products must be considered. Core definition teams will get meaningful input about them only if there is a product roadmap that stretches several years into the future. Since the responsibility for product roadmaps resides within product marketing or definition, it becomes apparent that the core definition phase involves a great deal of communication with several functional units (marketing, definition, development) within possibly many business units.

The essential tasks in the core definition phase of the Core Supply Process are to

- collect mostly company-internal customer input to identify common core requirements that enable multiple core usage
- check external core providers, compare offers, and perform acceptance checks
- estimate core development, support, and maintenance effort
- estimate profitability of core development/core purchase versus internal development of blocks that are not planned to be reused
- prepare core project plan for core development/purchase.

At the end of this first phase we have a milestone "core project release" where the responsible core working group decides whether to set up a core development project or not. The major decision criteria are the profitability of core and the expected core usage fee.

From this discussion one might believe that core definition teams could specify the right core functionalities if they just get all the requirements for using these cores in various applications. However, core definition cannot be purely reactive to the business units' requirements since diverging application areas will severely limit the degree of reuse of a core. Maximal reuse can be achieved if product roadmaps of different business units are aligned such that they need common functionality to be realized as cores. Financial model computations [Keat98a] clearly illustrate this point. The Core Supply Process provides a mechanism for such coordination among business units.

Core Realization Phase: In this phase the core is actually implemented. The first step is the detailed target specification of the core. We do not assume a strictly linear sequence of core specification, core design, and documentation steps. In practice, the core's specification will concurrently evolve with the core design up to a certain point ("specification freeze"). Thorough documentation of the core is essential for its successful reuse, and needs to be done early. Documentation tends to have low quality if written after the design is complete, provided that documentation is then written at all. The essential tasks in the core realization phase of the Core Supply Process are:

- create the core's target specification
- develop concepts for design, test, and verification
- design the core, and develop all views according to core quality standards
- review the core to check its compliance with core quality standards early in the project and again at the end
- create documentation for future core designers and core support, documentation for core integrators, a user manual for reference in product user manuals, and training material

At the end of this phase we have a milestone "core release" at which the core working group reviews the status of the core with respect to initial and current requirements. Also, the core owner responsible for core support and maintenance is defined.

Core Integration Phas: After development of the core, its availability is made known to product definition and development teams through the core catalogue. The core library, on the other hand, gives the designers access to the core deliverables. Access control need only be implemented for the core library. The essential tasks in the core integration phase of the Core Supply Process are:

- put the core into the core library (views and documentation) and the core catalogue (basic technical information and usage experience)
- support core integration in pilot project
- perform core integration review (with integration engineer and core developer)
- verify core on silicon, evaluate core's manufacturability
- core maintenance (new CAD views, bug fixes)

At the end of this phase, the core has been integrated already once, it has been verified on silicon, and is being maintained for and can be supported during further integrations.

To put this process in perspective once again, consider the reuse financial model study by Mike Keating [Keat98]. The study assumes 5-10 uses of an IP block per year on average. This cannot be achieved unless all activities related to IP definition, design and use are organized using a well-thought-out and well-documented process. To ensure the rapid introduction of the CSP into a semiconductor company, supporting actions concerning the corporate organization and intra-company business models are necessary.

4.4 ORGANIZATION

Besides requiring an infrastructure for core-based design, the CSP mandates new tasks in the definition, development, support and maintenance of cores. The business units' product teams are under severe pressure to get products out. Therefore, dedicated resources are required. This is essential. Earlier approaches considered requiring product designers to develop reusable functionality. This usually does not work, because of the time pressure to design the product. The designers are torn between apparently conflicting goals: bring a product to market soon, and define, design and document some functionality such that it is easily reusable for other products. Designers will choose to focus on their primary goal: get a product to market. Reusability will suffer. The question now arises whether to establish these resources in a centralized support organization or decentrally in the various business units (BUs).

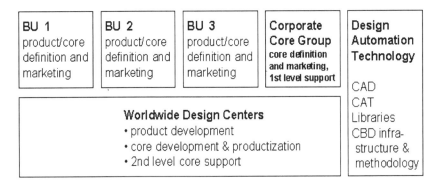

Figure 4.2 Organisation

There are advantages from having these dedicated resources located decentrally in the BUs: being close to the market, the BUs have the competence to identify and define application-specific cores. Therefore, their development teams also have the competence to develop and support such application-specific cores. However, there will be a conflict between product and core development, as product development leads faster to revenues. Also, BU core teams can only support core integrators in other BUs if an appropriate intra-company business model provides adequate financial rewards for their efforts. Even then, each BU will likely prioritize their own requirements in defining a core and thus lower the reusability of cores across BUs. A central group does not have these drawbacks, however, for a large company a central group is too far removed from the various markets to adequately address the requirements of the various BUs.

Figure 4.2 presents a working solution to this dilemma. The BUs focus on product definition and marketing, and also handle core definition for the cores that are specific to their business. A corporate core group handles core definition, marketing and 1st level support for standard and general purpose cores. The design centers all over the world are responsible for product development as well as core development and core productization. Increasingly, there are dedicated groups within the design centers that exclusively focus on core design. The design centers also provide the specialized 2nd level support for the cores that they own. Basic infrastructure for reuse, e.g., methodology, design flows, library infrastructure etc. are provided from the central Design Automation Technology group.

4.5 BUSINESS MODELS

There has been widespread discussion about the right business model for the exchange of IP between companies. Many independent IP providers favour a royalty-based business model, where a customer pays a fixed percentage of revenue, e.g. 2%, for a single core usage. This model is clearly not practical for future scenarios with dozens of cores on a single chip; royalties would easily exceed the average gross margin. Such a model does not reflect the fact that a single core's contribution to the total chip area is decreasing as chip complexities grow over time.

Not only business models for IP exchange between companies, but also business models for exchange of cores within the same corporation are required. The latter issue is the more immediate challenge for most large corporations that want to exploit their existing IP. Setting up such a business model is one of the most challenging aspects of a corporate design reuse strategy since it must reconcile several diverging goals.

An intra-company business model must provide incentives to business units to develop and support cores. Instead of spending additional time to develop cores that can be reused by other business units, engineers could develop new products. Therefore, core creation must be profitable for business units to justify the extra effort; a refund of the additional development costs is often not sufficient for the business unit. This profit required by development teams in business units increases core usage fees. Once cores have been developed and are available, however, core usage fees should be as low as possible to foster reuse and to get the best return on the investment made into the core. Moreover, the business model must be designed to finance core development within business units as well as corporate core groups. Corporate core groups, however, are not required to make a profit; similar to other corporate technology groups, they need funding for their work. Furthermore, the business model should be easy to administrate, while accurately reflecting actual costs.

We try to reconcile these requirements by a business model that is based on the sum S of the actual core development cost, the expected basic support costs, and a fixed percentage of profit in case the development and support is done by a business unit. At the end of the core definition phase ("core project release"), there is a number N of customers, who commit to pay an even share S/N of S. This is the initial core usage fee. The core development group will achieve additional revenue from follow-up customers. Since they may be charged a usage fee larger than S/N, there is an incentive for each business unit to become a core customer and to contribute to the core definition early on.

4.6 ENSURING CORE QUALITY

"The reuse battle is won or lost based on the quality of the IP" [Keat98b]. There are two different perspectives to Core Quality. First, a core must have competitive area, performance, and power consumption. These must be captured in the core requirements during the core definition phase of the Core Supply Process. Second, a core must be prepared for easy reuse. Requirements for easy reuse are captured in the *Core Quality Guidelines* of Siemens Semiconductor.

The Core Quality Guidelines describe mandatory and optional core deliverables (such as CAD views and scripts), documentation requirements, HDL coding guidelines, and design and test guidelines. Explaining and motivating the Core Quality Guidelines is an important part of the reuse training that every designer should go through. To make sure that a core is compliant to the Core Quality Guidelines, core reviews are done by the corporate core group with the core developer. Core integration reviews are done with chip integrators to capture experience gained during core integration, and to provide feedback to the core creator about improvement potential. Since manual reviews require significant effort, they cannot be applied to every core, and they cannot be applied early and often enough. We are developing and integrating automated checking tools into our design flows to simplify manual checking as much as possible. Commercial tools are currently available only for automated HDL checking.

4.7 TECHNICAL ISSUES

Technical considerations of design reuse have been described in much detail in a large number of publications recently. Here, we will mention just the major aspects of Siemens Semiconductor's design reuse strategy without going into detail.

Design Flows and their component **EDA tools** need to adequately address design reuse. Both the creation and the integration of cores need to be addressed. It is not necessary to entirely introduce new flows, but to modify and enhance existing flows. At Siemens Semiconductor, Semicustom Highway and Fullcustom Freeway are highly automated design environment with state-of-the-art CAD tools, for language-based digital design and schematic-based fullcustom/analog design, respectively.

A key enhancement of existing design flows is the ability to efficiently perform electrical characterization (timing, power) of cores. The obtained characterization results then need to be efficiently modelled for the various tools in the design flow. In the Siemens Semiconductor design flow, the Freeway2Highway Interface assumes these tasks. The effort for this task would be much reduced if standardization attempts (e.g. OMI, OLA, ALF, LIBerty) would finally succeed. For synthesizable soft cores, the handling of configurations, online-access to documentation, library-specific and technology-specific adaptation of scripts (synthesis, STA, scan insertion) can be much improved by packaging cores, e.g. supplying them with a graphical user interface. Modelling environments and guidelines for C-modelling and simulation are required both internally (used as executable specifications) and externally (used as early customer interfaces). Verification methodology is addressed by defining standard testbench concepts, making module-level testbenches partially reusable at the chip integration level, by providing HW/SW cosimulation environments etc. Production testing should not turn into a new bottleneck, therefore, improved design-for-testability measures need to be introduced, e.g. logic BIST, use of test busses, automated integration of module-level testpattern at chip-level.

Today's commercial tools typically do not adequately address these requirements. Siemens Semiconductor addresses them both by developing internal solutions where appropriate. In other cases, we engage in early cooperations with key EDA vendors. This allows us to work with them to guide their R&D investments to maximum benefit and also to prepare our design flows at a very early stage for integration of new tools.

Design methodologies likely will also see some changes, but only in a second phase. Physical considerations cannot be contained in the flow backend. They will move up in the design flow. Early RTL-based floorplanning incorporating existing cores will become state-of-the-art. Top-level design planning (topology planning, but also timing budgeting) will be of more importance than today.

The **design of reusable cores** requires significant attention to reusability considerations. Configurability ensures their adaptability to different design projects. Typical configuration parameters are bit widths of busses, addresses of memory-mapped registers, selection/deselection of features. A standardized on-chip-bus[1] enables communication between instruction set processors and their peripherals while freeing the designer from the effort of designing application-specific communication structures. A supportive environment needs to be set up for the bus, with

---

1. While VSI-A did not succeed in defining a single (or a few) on-chip-bus, within one company this is a more reasonable goal. Nevertheless, to allow e.g. peripherals to be used with multiple bus interfaces, we design them with a generalized point-to-point interface. This is supplemented by adaptors for the different supported busses. This approach is similar to the one that VSI-A adopted later on.

protocol checkers, compliance validation suites, electrical characterization of the bus and a library of basic building blocks, such as master and slave implementations and bus controllers. Other standards relate to clock architecture, power architecture on-chip debug support interfaces etc. In general, standards are important to ensure success of design reuse, especially when cores are exchanged between companies. This is the reason for the importance of the work of the Virtual Socket Interface Alliance.

A **core library** including version control and easy-to-use delivery mechanisms also is essential. Earlier, cores used to be delivered via E-Mail, one mail for each deliverable, without version control etc. This is not sustainable. Library mechanisms also need to support protection of cores. The core library needs to be supported by a core database that provides rapid information about available cores, their key technical data, their previous uses etc.

Finally, **training** is essential to ensure that designers get maximum possible benefits from new methodologies, flows, tools and supporting infrastructure.

4.8 THE OVERALL STRATEGY

As part of its Logic Initiative, Siemens Semiconductor has established the Core-Based Design reuse strategy. It consists of a coherent set of technical initiatives as well as non-technical process and organizational approaches. The foremost non-technical aspect is the Core Supply Process, which guides the lifecycle of a core from idea through definition and design into the support and maintenance phase. The CSP is supported by an appropriate corporate organization as well as internal business models. Metrics ensure that the progress that is achieved is continuously monitored.

These non-technical aspects of our strategy are complemented by very aggressive investments in design methodology and the required EDA tools. Cooperations with leading EDA vendors support our in-house efforts. Best-in-class design techniques are determined and defined for the core and product development groups. To ensure the future growth of the industry, Siemens Semiconductor actively participates in standards committees such as VSI.

5 DESIGN METHODOLOGY FOR IP PROVIDERS

J. Haase, T. Oberthür and M. Oberwestberg

SICAN GmbH
Hannover, Germany

5.1 HOW TO BECOME AN IP VENDOR

Meanwhile it is accepted as a fact by most designers that for developing Systems-on-Chip comprising tens of millions of gates, in the near future the application of reuse-methodology is mandatory. Already today's application of reuse pays off in terms of development cost and time-to-market.

As first conclusion most system companies start programs to ensure reuse of the modules designed in-house. On the other hand many companies including design houses, tool vendors and a lot of new start-ups have the vision to become a vendor of Intellectual Property (IP).

SICAN as one of the leading companies in the IP business started very early to provide its know-how to customers in form of reusable soft cores. They provided their DesignObjects™ (SICAN's trademark for IP) to a large number of users worldwide. Thus SICAN has accumulated substantial experience, technical as well as commercial, in this new market. Based on that experience this paper wants to present some prerequisites and solutions for being a successful IP vendor.

As a basic rule an IP vendor has to be aware of the effort required for developing DesignObjects™ (in the following DesignObjects™ shall mean reusable soft cores coded in VHDL or Verilog-HDL). It is not sufficient to take modules from a larger design and to isolate them. SICAN's experience in numerous designs show that designing real DesignObjects™ takes about 2 - 2.5 times the effort of a regular design. Typically, this extra-effort pays off when the DesignObject™ is reused at a minimum three times. Reduced time-to-market resulting from the use of pre-designed IP and reduced development cost will be the benefits for the customer, not for the IP vendor himself. The IP vendor has to make sure that he will be compensated for the extra-effort by selling the DesignObject™.

Some of the key issues that need to be addressed by an IP vendor during the development phase are:

- marketing research
- test environment
- design automation
- support of differing design platforms
- scalability
- parametrization
- product documentation

The following chapters describe an approach gained from practical experience in reuse projects.

5.2 IP DATABASE STRUCTURE

In order to successfully handle design projects, it is necessary to have a well defined data structure. This is even more important during IP development and distribution. The database consists of the directory tree, the location of files (like HDL source files, verification pattern, scripts, setups), file naming styles, the setting and use of environment variables etc. Figure 5.1 shows an example how the directory structure of each module in a design can be defined.

The directory tree and all data files should be managed under a revision control system. This ensures a safety development process and it is a key prerequisite for a DesignObject™ delivery system to manage different releases, tracking bugs and to ensure data consistency (cf. Chapter 5.5).

The use of a defined naming style for at least all source files of the design simplifies the use of general scripts to manage the database. It enables each designer and later on the customer to easily navigate through the design database.

General scripts and setup files based on the structure and naming style can be placed at a global position in the database structure. They can be used e.g. for software environments, synthesis and process automation.

Based on a synthesis script system the mapping of a DesignObjects™ to a target technology can be performed. The global approach simplifies the adaptation of synthesis scripts to different technologies.

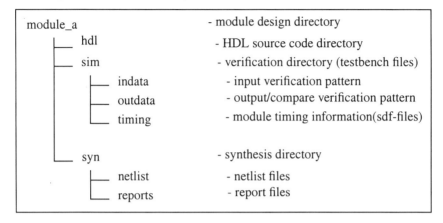

Figure 5.1 Example of a directory structure

For process automation the use of the make utility guarantees for an easy way to recompile large designs for synthesis and simulation after modification of the HDL source code. The use of global include files simplifies the adaptation of the makefiles to different environments.

All general scripts should use UNIX environment variables to define paths or settings, differing from customer to customer. This allows for providing unified installation procedures for DesignObjects™, required for automation of the installation process of the DesignObjects™ database at customers site. Finally, the clear structure of the design database eases the integration of the DesignObjects™ sources into the customers' environment.

5.3 DOCUMENTATION OF IP

One of the most important issues of reuse is a good documentation. The documentation should contain all information the customer needs in order to handle the DesignObjects™ in his environment and to integrate them in his system. First of all a set of documents with an uniform format needs to be defined, which is provided with each DesignObject™. Two different kinds of information can be distinguished. A functional part describes the functionality, architecture and interfaces of a DesignObject™. An implementation part describes the data structure, the way the DesignObject™ was designed and how it can be verified and implemented in the customers' system environment.

Based on this kind of information, two different types of DesignObject™ documents can be defined:

(1) "Black Box"-Manual:
Used for DesignObjects™, which can be delivered "off-the-shelf"

(2) "White Box"-Manual:
Used for DesignObjects™, which require additional design effort for adaptation to the system requirements of the customer. It has to include additional information about the internal structure of the DesignObject™.

As an example the content of a "Black Box"-Manual can be defined in the following way:

(1) Features and Functional Overview
Describes the main features of the DesignObject™ and gives a short functional overview.

(2) Interface Description
Lists all signals and accessible registers and describes their function and possible settings. Provides timing diagrams to describe the signal handshake

(3) Programming Guide
Describes the operation modes and functions of the DesignObject™ and how it can be programmed or configured.

(4) Data Structure
Provides information about the DesignObject™ database. This includes all information, which is necessary to handle the database (e.g. directory structure, design structure, setups, etc.)

(5) Behavioural Model
Gives all information that are necessary to understand, compile and use a behavioural model of the DesignObject™.

(6) Implementation Information
Provides information about the design and synthesis methodology. This includes information about clock tree, internal RAMs, asynchronous design parts etc. and how the DesignObject™ can be mapped to a target technology. Additional information is given to characterize the design in terms of frequency, size and power.

(7) Verification Information
Describes the methodology which is used to verify the design. Gives information about the testcases and how new testcases can be created.

(8) Test Information
Gives information about the implemented test structures.

The key to reusability is not only the transfer of reusable code, but the transfer of the knowledge to handle the functionality therein. The quality of the documentation cannot be overestimated and will be one of the key characteristics for an IP provider.

5.4 SIMULATION TESTBENCH PHILOSOPHY

Design verification can be described as a process, in which the designer determines whether the design operates correctly according to the specification. A testbench can be defined as a system to verify the design during simulation. This testbench can be very simple by using defined testvectors or very complex by using a behavioural model to check the functionality of the design. In VHDL-based simulation systems, a testbench generally consists of a design instance (Unit Under Test - UUT), processes and functions as well as models of components which will interact with the design. The testbench generates stimuli that exercise the design and verify the correct functionality by monitoring the design's output ports.

Figure 5.2 Top level testbench generator window

For DesignObjects™ it is very useful to apply a universal concept for all testbenches. First of all, an uniform testbench structure should to be used. This can be achieved by using templates generated by a script or program. Figure 5.2 shows the top level window of a testbench template generator written in Perl/Tk.

In the first section a HDL search path is defined. This path is the starting point for searching in any design files. The second path specifies the directory where the generated template files are to be stored. In the next section the name of the UUT is specified. Furthermore, the clock and reset signal names and a prefix for instances can be defined. In the third section, additional modules, that should be placed in the testbench and VHDL packages, can be defined. In the fourth section some options can be specified and the template generation process can be started.

Four VHDL files are generated: an entity file, a configuration file, a package file and an architecture file. The script searches for the HDL source files of the UUT and any other modules, instantiates the modules in the architecture file and automatically does the signal definitions and module wiring. Additional global procedures for clock and reset generation are instantiated.

Furthermore, a template for a verification information file called "vplan" is generated. This file is used to describe the whole verification process. Four main chapters are used for description:

(1) Verification Definition
This chapter defines the kind of tests, that need to be performed to verify the module. This includes tests to verify the function and the interface behaviour

(2) Verification Structure
The HDL files used for the testbench are explained. Furthermore, a brief description of the processes, procedures and functions used by the testbench is given. Finally a short description of the stimulus process should be given

(3) Test Data Environment and Generation
Format descriptions of all test data files, that are used during verification process. Any programs or scripts used for test data generation are explained as well as the way how new tests can be specified

(4) Verification Invocation
Describes the setup of the verification environment and how tests can be started.

An example of a general testbench structure is shown in Figure 5.3. The stimulus process can generate log messages and waveforms for the UUT. The waveforms can be generated by the stimulus process itself or by starting generator processes. The generator processes are used for data (waveform) generation, for observing the outputs of the UUT and for automatically checking the results. Three different packages are used to define the generator and observer:

(1) Global package: used for all module testbenches of the DesignObject™ like clock and reset generator

(2) Library package: used for at least two module testbenches of the DesignObject™

(3) Module package: package for one module with generators and observers that can't be shared

This approach leads to a structured concept for all testbenches of a DesignObject™ and helps the customer to understand the verification flow. Furthermore, it speeds up the development process for the IP provider by reusing procedures and functions.

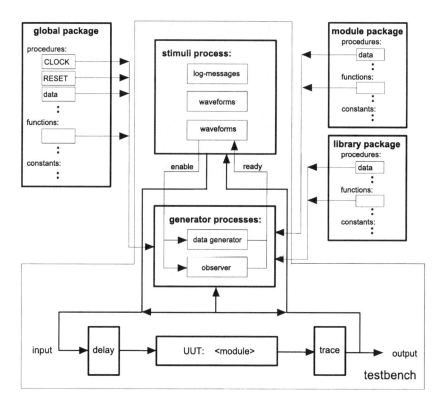

Figure 5.3 General testbench structure

5.5 RELEASE MANAGEMENT

IP is developed under the aspect of reusability. For an IP provider this implies that the macro cell will be delivered to more than one customer at different times. Keeping track of different releases with possibly slightly modified functionality is one the major tasks an IP provider faces apart from the design itself.

Each delivery of a DesignObject™ to one or several customers is a separate release. The key tool for an efficient release management is an advanced revision control system which keeps all files and information related to an DesignObject™ in a single repository. The revision control system has to provide the means to

- store previous versions of a file
- track the history of a file
- label a certain state of the design (a release) with a meaningful name
- extract previous releases based on labels
- allow creation of isolated branches
- support merging changes in branches back to the main tree of development

All these requirements are essential for handling data management, bug tracking and general maintenance.

5.6 DUAL LANGUAGE DESIGNOBJECTS™

The digital ASIC design world is divided by two major hardware description languages, VHDL and Verilog-HDL. While Verilog is mainly used at the U.S. west coast, VHDL is preferred in most other parts of the world. Companies using IP from different vendors usually do not want to assemble their design with blocks described in different languages. Mixing VHDL and Verilog in one simulation is supported by most commercial simulators, but nevertheless there are some drawbacks from using this feature. Having a mixed language design additionally requires the designers to be fluent with both languages which requires quite an amount of experience.

For globally acting IP providers, this dilemma means that it therefore is best to provide their IP in both languages. This causes additional effort for implementation, verification, maintaining a consistent database, bug tracking and fixing as well as release management. Reducing this effort is a critical issue for IP providers.

The worst solution for this problem is to have two teams working in parallel, one using Verilog the other one coding in VHDL and both implementing the same functionality. The effort required for initial implementation and verification is twice the effort for supporting only one language. Post implementation effort like bug fixing or enhancing functionality is also a linear function of the number of supported languages. The double effort causes costs which are usually not compensated by additional income through selling the DesignObject™ twice as often.

A better solution to reduce the additional effort for supporting two HDLs is to choose one HDL as the master language and automatically generate the second one with a translator. This solution requires an extended design methodology which integrates an automatic translator into the flow. This concept also requires a project setup in which it is easily possible to switch between the two versions of the DesignObject™ to reduce the overhead of verifying both versions.

VHDL supports more elaborated language constructs than Verilog. Therefore if VHDL is chosen as the master language only the subset of the language may be used which is supported by the translator. This subset varies between different translators. Thus a coding guideline should be developed and maintained as an ongoing process. The following paragraphs assume VHDL as the master language.

Basically the flow for automatic translation consists of three steps as shown in Figure 5.4.

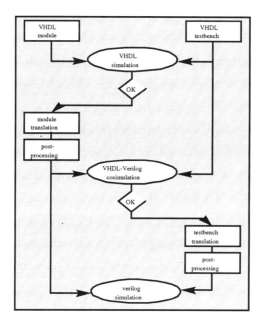

Figure 5.4 Example of translation flow

As the first step the synthesizable VHDL design is translated to Verilog. During translation, language constructs unsupported by the translator are revealed. The VHDL sources are modified until the code passes the translator. In the next step the generated Verilog code is co-simulated in the VHDL testbench environment. These co-simulations are required to detect problems which are due to different simulation approaches of both HDLs. Examples are the use of blocking and non-blocking assignments in Verilog and the different handling of default assignments.

Once the co-simulation is successful, the VHDL testbench has to be translated. After this step, pure Verilog simulations can be performed as last step. The translation of synthesizable code usually is no problem as only a relatively small subset of VHDL is supported by most synthesis tools anyway. This subset is generally also available in Verilog. The situation is different for testbenches which often use more behavioural language constructs. Therefore, writing a translatable VHDL testbench requires more caution and poses more limitations on the designer to be considered.

One approach for keeping testbenches translatable is to move the task of generating test and reference data from the testbench to external models. Feeding the simulation with test data and comparing the simulation results with the expected values can then be handled via simple file I/O and compare functions.

As VHDL and Verilog differ regarding file I/O functionality, Verilog might have to be extended via its standardized programming language interface (PLI).

Once a steady flow for automatic translation is established like it is shown in Figure 5.4 and a coding style guide is available for the chosen translator the task of generating one HDL from the other is quite straight forward and the co-simulation step can be skipped.

Generating new DesignObjects™ in both HDLs requires only a minimum overhead compared to generating DesignObjects™ for only one language. Maintenance is simplified because bug fixes and enhancements only have to be applied to the master source code. With an automatic translation step all changes are automatically passed to the second HDL version of the DesignObject™. Makefiles or other dependency checkers can be used to ensure a consistent source code data base.

5.7 SCALABLE DESIGNOBJECTS™

Another important challenge for an IP provider is to match the customers' needs as perfect as possible. In most digital designs, gate count still is a major concern. Having DesignObjects™ providing functionality the customer doesn't really need at the cost of a higher gate count, this might prevent the customer from using that DesignObjects™.

Example: A semiconductor company plans to design a chip for DVD applications containing audio and video decoding. Depending on the target market AC-3 (e.g. in North America) or MPEG 2 (e.g. in Europe) need to be supported. The semiconductor company might want to have dedicated chips for each region or have a single chip solution supporting both audio decoding algorithms or even all three versions. The semiconductor company wants to buy the audio decoding part of the chip from an IP provider.

Both audio algorithms are based on similar theoretical background. Having a combined decoder supporting both algorithms can benefit from resource sharing to reduce gate count. Nevertheless, if only a single algorithm needs to be supported that solution would be oversized and a dedicated decoder block would fit the customers need better.

The IP provider faces a similar problem as already described in the previous chapter for supporting more than one HDL. The best solution for the IP provider regarding effort and costs for implementing, verifying and maintaining the DesignObjects™ would be to have one source code base and derive different versions of the DesignObject™ from that source code base automatically.

Such parametric or scalable DesignObjects™ provide the means to tailor their functionality to what the customer really requires. Unfortunately, neither VHDL nor Verilog include this as part of the language definition. Although VHDL and Verilog offer methods to pass parameters or generics to a module to scale certain values like data path width, this kind of scalability does not allow to manipulate the overall architecture or functionality of a module.

A solution to extend the VHDL and Verilog functionality is to use C-like preprocessing to generate the desired derivative from the source code base. The source files have to be extended by certain keywords to mark areas of the code belonging to a certain part of the functionality of the DesignObject™. A preprocessor removes these lines from the code base and extracts only the VHDL or Verilog code indicated by the setting of the preprocessor directives.

5.8 EXPERIENCE FROM REUSE PROJECTS

At SICAN a family of audio decoder DesignObjects™ was designed following the methodology described in the previous two chapters. This family consists of a combined AC-3/MPEG 2 audio decoder, a MPEG 2 only and an AC-3 only decoder. All versions are available in VHDL and Verilog.

VHDL was chosen as the master language for the DesignObject™ family. The architecture was developed with encapsulation of a specific algorithm and common parts of the architecture in mind. During implementation these code fragments were already enclosed with preprocessor directives. A preprocessor was build to extract the required code for a given version of the audio decoder. Scripts were used to support the generation of a decoder version from the source code base with just one program call.

Verification is based on file I/O with stimuli and reference data generated from C-models which results in a HDL independent verification environment. There are two separate models for AC-3 and MPEG 2. The format of the generated files is defined equally for both algorithms so that the testbench itself does not have to take care about the simulated algorithm. The testbench environment is set up using the generic approach shown in Figure 5.5 which makes it easy to be reused for other designs with translation from VHDL to Verilog. The design under test (DUT) is surrounded by stub modules which handle feeding test data into the design and to request and compare the simulation output with the expected results. There are dedicated stubs for each interface of the DUT. Due to the equal structure of all used files all stub modules can refer to the same set of functions for reading and processing stimuli and reference data. The testbench itself is self-checking which enables an automatic verification flow. All required tests are stored in a test database. Certain critical tests are tagged as regression tests.

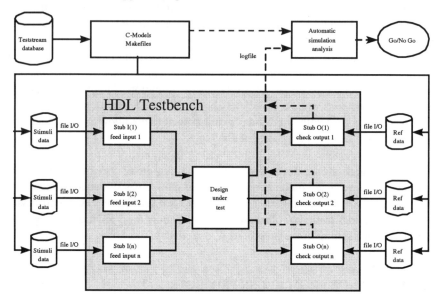

Figure 5.5 Testbench environment

After code modification it is possible to run regression or full tests without any user interaction. Reports are automatically generated indicating success of failure of the executed simulations. The VHDL code is fed into the VHDL to Verilog translator to automatically generate the Verilog version of the design.

As this project was the first project using the automatic language translation flow, VHDL to Verilog translation was introduced early in the implementation phase to determine translation problems and build up a coding guideline for the selected translator. Finally, VHDL was written in a way which does not require manual postprocessing of the generated Verilog code. Due to the regular architecture of the testbenches, a relatively small library of Verilog file I/O functions was implemented using PLI. Therefore, the testbench structure shown above could also be translated without any problems.

As a result a whole family of the DesignObjects™ based on one single source code base could be developed without much overhead compared to just a single combined AC-3/MPEG 2 audio decoder. Table 5.1 shows the gate count statistics for the decoder family

Table 5.2 lists the effort spent for developing the decoder family. The overhead for translation is dominated by determining the appropriate coding style for the selected VHDL to Verilog translator. This number will be much less once a good coding style guide is available. .

Version	gate count
AC-3/MPEG2	68.000
AC-3 only	51.000
MPEG 2 only	36.000

Table 5.1 Gate count for decoder family

All files to be delivered to customers are under control of a revision control system. In case of a delivery, all files are tagged with a symbolic name representing the version being delivered. Branches are generated for each customer in which the environment of the customer is mirrored.

	Effort
Design effort (reusable)	85%
Overhead for scalability	~ 5%
Overhead for translation	~ 10%

Table 5.2 Effort for decoder development

5.9 CONCLUSIONS

This contribution presented some basic methodology suitable for successful development and reuse of IP. It covers general design methodology, deliverables for DesignObjects™ as well as technical solutions for dual language and scalable DesignObjects™. This was completed by experience and results gained in commercial reuse projects.

Besides developing its own reuse methodology an IP vendor has to rely on standards. The VSIA (Virtual Socket Interface Alliance) does a lot of important work on this. However, by providing standards for the technical issues only one part of the problems are solved. Making reuse a commercial success requires the development of adequate business models for selling and applying IP. It will take some time until all standard business models become accepted on a worldwide basis. After all, this is important not only for IP vendors and their customers, but also for making in-house reuse a reality. When a company establishes an internal IP-pool, it also needs internal business models to compensate its design teams for contributions to this pool and for the effort to achieve reusability. Without such compensation, the extra-effort for designing really reusable modules cannot be spent.

 **Jürgen A. Haase** received both his M.S. (1983) and Ph.D. (1989) degrees in Electrical Engineering from the University of Stuttgart, Germany. From 1989 to 1995 he worked at Philips Communications Industries. Currently he is Design Factory Manager at SICAN GmbH, Germany and is responsible for SICAN's design services and IP products. His research interests include applications of digital signal processing, design of systems-on-chip and reuse methodology.

6 HARD IP REUSE METHODOLOGY FOR EMBEDDED CORES

W. Eisenmann, S. Scharfenberg, D. Seidler,
J. Geishauser, H. Ranise and P. Schindler

Motorola GmbH
System on Chip Design Technology
Munich, Germany

6.1 INTRODUCTION

Today's hierarchical cell based design methodologies, tools and systems are capable to handle soft IP fairly well by treating it just like another RTL block. Improvements are desirable in the areas of synthesis constraints and ease of use, but the basic capabilities required to design a chip including soft IP do exist. This is almost true also for hard IP, if the block is treated like a big standard cell. Modifications to the design system are only necessary for IP protection purposes and embedded core test support. To support the standard cell approach with hard IP the library views for all tools used in the design flow need to be created. Figure 6.1 shows the hard IP deliverables that are supported today indicating for which of the views we have automation in place and which are still generated manually.

64   REUSE TECHNIQUES FOR VLSI DESIGN

This paper covers the hard IP view generation and reuse methodology that has been developed and successfully applied to most of the key Motorola embedded cores, like the 68K, ColdFire™, M·Core™ and PowerPC™. The same processes can and have been applied to non-programmable circuit blocks designed with cell based or full custom methods. We have proven that a cell-based flow is sufficient and in fact the fastest way to put together a system on a chip.

Special focus will be on the following five areas: simulation models, timing characterization, embedded core test, frontend and backend views, IP storage and distribution. IP protection issues and required design system enhancements will be discussed in all five areas.

Tool	Format	Automation
Synthesis	.db	√
Delay Calculation	proprietary	√
PLI wrapper	Verilog /QHDL	√
Timing Wrapper	Verilog /QHDL	√
Bus Wrapper	Verilog /QHDL	√
FFM	object code	√
BFM	Verilog	manual
Static TA	.mml	√
Floorplanning	proprietary	√
ATPG	fastscan model	manual
Cell3/Silicon Ensemble	LEF	semi automatic
Aquarius/Apollo	binary db	semi automatic
Layout Verification	CIR, GDSII	design system + manual îhardeningî
test pattern	WGL	design system
documentation	PDF	manual

Figure 6.1 Hard IP deliverables

6.2 SIMULATION MODEL GENERATION

Several functional simulation models are required for a hard macro core. A bus functional model (BFM) is generated manually based on the bus specification. It is usually written in C or Verilog/VHDL and drives the simulation with the core's bus response. The BFM is mostly used during development of peripherals.

For full-chip simulation and test pattern re-simulation one or more full functional models (FFM) are required also. Due to IP protection reasons the RTL or gate level netlist can not be used. Therefore the netlists need to be translated to C code, compiled and linked into a shared library using the Verilog Model Compiler from Synopsys. Verilog wrapper modules include the PLI-call and a timing shell for the core.

HARD IP REUSE METHODOLOGY FOR EMBEDDED CORES   65

The design system has been enhanced to automatically loop through all hard IP a designer selected for his chip, reading the Verilog wrappers and adding the shared library to the LD_LIBRARY_PATH environment variable. The user can select between Verilog, VCS, QHDL and Leapfrog simulators which are all supported in the design system on Solaris and HP10 platforms.

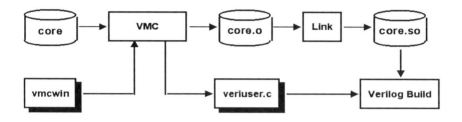

Figure 6.2 VMC model generation and simulator build

6.3 STARTERKIT SIMULATION ENVIRONMENT

The StarterKit is a core based simulation environment that enables the user to do a software driven simulation very easily. It consists of the core, a virtual memory model and setup/clock generation blocks. The StarterKit is not only meant to be used as the starting point for a chip design, but it helps to speed-up the learning curve on how to use a specific embedded processor.

The internal clock generator and the reset and setup logic is coded in RTL. The core is a protected VMC model. VMC is a program that compiles the Verilog code to C-code. This compiled C consists of a small Verilog simulation engine that communicates with the master simulator via a PLI interface. This type of model has the advantage that it is an exact model of the core, but has the disadvantage of a slower simulation speed compared to an Instruction Set Simulator (ISS).

The memory model is a mixture of a hand coded C-Model and a Verilog wrapper. Figure 6.3 shows the different modelling layers in the communication between the Core and the Virtual Memory model. The communication of a read operation for example is initiated from the VMC core model that passes the PLI wrapper, which connects the ports of the VMC-Model with the Verilog world. In addition to that the driver strength modelling is done here. After passing the PLI wrapper, the transfer passes the timing wrapper which was introduced to be able to do back annotation of timing to the core model (see [Scha98]). The connection block represents the wiring between the Core and the Memory Model. The last instance that the transfer has to pass is the Verilog wrapper which transforms the Core bus protocol to the Virtual Memory(VM) port protocol. This means that only this wrapper has to be changed whenever the VM should be connected to a new core.

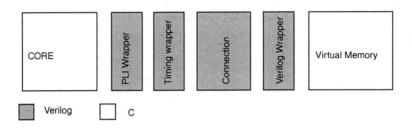

Figure 6.3 Modelling layers

The reference structure design (see Figure 6.4) differs form the minimal core design in the way that it includes also the reference bus hierarchy. This hierarchical bus design was generated to have different plug in possibilities for peripherals, depending on their data transfer requirements.

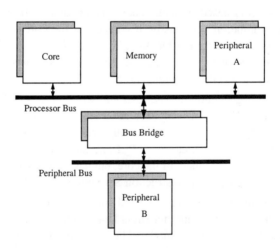

Figure 6.4 Reference structure design

In the example two peripherals are plugged into the design. Peripheral A is connected to the processor bus, which means that it can benefit from the high bandwidth, but on the other side it has to take care of the complex protocol and strong timing constraints. Peripheral B is plugged to the peripheral bus.

If a processor based design has to be done, there is always some kind of software that has to be written. Therefore a free software cross development package is included in every StarterKit. The basic idea behind software driven simulation is that a processor executes opcode and accesses models written in Verilog to prove their function. In contrast to this, the classic approach uses a testbench written in Verilog/VHDL to generate the required stimulus. There is still some Verilog/VHDL code in the StarterKit that acts as a testbench, for example some code that handles the serial signals of a peripheral, but it is much less and easier to maintain.

The design flow of the StarterKit is shown in Figure 6.5. The program code is getting assembled and converted into an S-RECORD format file. The S-RECORD format is the interface between the software development and the hardware simulation environment. This also guarantees that the user can exchange the software tool chain since almost every software development package is able to generate S-RECORD files. This file is loaded by a C-Routine into the Verilog simulation environment. After running the simulation, the execution of the program can be displayed by a waveform viewer. See [Geis98] for more information on the StarterKit.

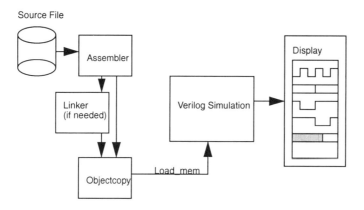

Figure 6.5 Data flow for a software driven simulation

6.4 TIMING CHARACTERIZATION AND TIMING MODELS

Besides the functional behaviour of a core its timing characteristics need to be captured also. Timing characterization plays a central role in the view generation process, because the results are needed for a number of different purposes:

- timing data for digital timing simulation
- timing data for synthesis of the surrounding logic
- timing data for static timing analysis of the whole chip
- timing data for timing driven place & route

Due to the complexity and size of the circuits (100K to 1000K) it is impossible to measure all the timing paths through transistor level simulation. Therefore, our approach is based on static timing analysis. The central timing analysis tool is Path-Mill from Synopsys EPIC Technology Group. The tool runs on a transistor level netlist representing the final layout. Together with this Spice netlist, parasitic capacities or parasitic capacities and resistances that are extracted from the final layout of the module are used. Other inputs to the timing analysis tool are transistor technology data in a tabular format and configuration files for the tool itself. The results of a characterization run are visualized via PathMill's API interface. A set of API routines, programs and scripts has been developed that read the characterization results and store them in an intermediate timing database. Database clients can access the timing database and translate its content into other formats. Some additional post processing is also required for the final timing data. Figure 6.6 shows the timing characterization flow.

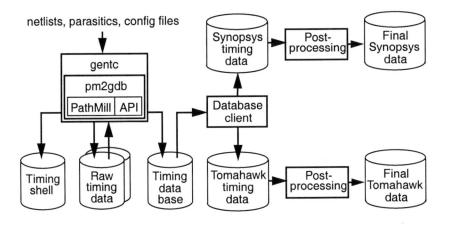

Figure 6.6 Timing characterization flow

One of the input criteria to this timing characterization flow is that a static timing analysis setup of the IP does exist. If the design group has already used PathMill during the design phase their configuration files can be reused in order to speed up the generation of the characterization setup. Once the correct setup is finished multiple PathMill runs have to be combined into groups with the same functionality. The goal is to create timing data and timing constraints for a black box model of the

IP. The four basic timing groups are shown in Figure 6.7. Combinational paths require the simplest setup, no clock tree recognition is required and only the delays of the longest paths are measured. In addition to that clock to output (clkout) runs require a correct clock tree recognition. The results are also delays of the longest paths. Tri-state runs can be combinational or clock related, depending on the design. Their results are again delays of longest paths. Setup and hold timing constraints are a result of investigating the longest and the shortest paths. Each of these timing groups requires their own special set of configuration files in which the IP is put into the right operation mode and the timing analysis tool is configured for the correct analysis.

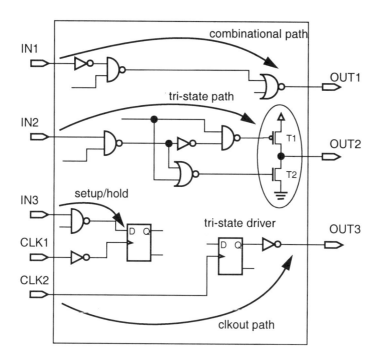

Figure 6.7 Black box timing groups

The delay and timing constraint numbers depend on the operating condition in which the module is used. They depend on the process corner, the voltage, the temperature, the slope of the input signals, and the capacitive load on the output pins of the module. The dependence of process corner, voltage, and temperature is covered by the transistor technology file for PathMill. The dependence on slope and load has to be covered by extra characterization runs.

Together with the module designer, the module integrator, and the requirements given by the design system it has to be agreed on a set of input slopes and output loads. The agreed set of n_clk_slopes different clock slopes, n_data_slopes different data slopes and n_loads different output loads specifies the environment in which the IP has to be characterized. Table 6.1 shows how the number of input slopes and output loads results in the number of required characterization runs.

Most static timing analysis tools do not understand the 'Z' high-impedance state, they can deal only with logical 0 and logical 1 states. A workaround is to model the 'Z' state as a function of the known '0' and '1' states. This can be done by identifying the tri-state drivers of the module and defining the gates of the driving transistors as sink nodes for the path search. The delay up to these driving transistors can be used for the tri-state transitions. The additional delay caused by the driving transistor itself has to be neglected. Tri-state slopes are difficult to measure, instead the logical '01' and '10' slopes are used. The timing data base access scripts take care that this is entered correctly into the data base as timing data of the tri-state pin.

	timing depends on			
timing group	clock slope	data slope	output load	number of characterization runs
combinational		x	x	n_data_slopes x n_loads
clkout	x		x	n_clock_slopes x n_loads
tri-state		x	x	n_data_slopes x n_loads
setup/hold	x	x		n_clock_slopes x n_data_slopes

Table 6.1 Number of required characterization runs

In order to do a full chip simulation with timing backannotation from a sdf file it is required to surround the simulation model of the module with a timing shell. The purpose of the timing shell or timing wrapper is to have variable and default timings for all the timing arcs and setup and hold constraints found during the characterization. It has also to handle the timing backannotation of bidirectional signals. In addition to that it has to provide extra input buffers for backannotation of interconnect delays. The timing shell should not add extra functionality to the module, but it has to provide a way to switch off setup and hold timing checks for example during the reset phase. It can also support pulse width checks for set and reset signals. Furthermore pulse width checks for the clock signal are added to make sure that the timing data is not used at a higher frequency than the module was designed for and has been characterized at. The timing shell is automatically generated by gentc (see Figure 6.6) using a Verilog netlist of the module and a configuration file.

The current approach of describing the timing of large modules is borrowed from the ASIC design style. All the tools and data formats were invented for smaller modules like standard cells. Some of today's tools and data formats have limitations which require several workarounds. Hopefully, future data formats like ALF or the new IEEE P1481 standard make this job easier.

6.5 FRONTEND VIEWS AND EMBEDDED CORE TEST METHODOLOGIES

A number of additional frontend views is required to support ERC, power estimation, floorplanning and static timing analysis tools for a design including some hard IP. We have created automatic translators for all these views, which are mainly Motorola proprietary formats, relying on existing input data. Most of the frontend views are derived from the Synopsys library files coming from the timing characterization.

The process corner dependant Synopsys sources containing capacitance and timing information are translated with the dc_shell Synopsys tool into binary files. Both the .lib sources and the .db binaries are part of the delivered core library. The Static Timing Analysis tool presently supported in the design system is Motive. The Motive core library file is generated by an encapsulation of the s2m (synopsys to motive) script. This library file is not corner dependent because we always require to read in an SDT file with the actual timing data before starting analysis. Power estimation is performed at the moment with a Motorola internal tool which requires a view for each corner, which is again generated from the Synopsys source. The ERC core view is generated from the pin information contained in the netlist by an internal generation script (gentc).

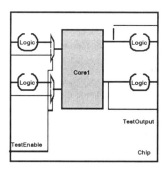
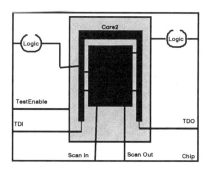

Figure 6.8 Multiplexing and internal boundary scan

The biggest problem in the frontend area is the test of an embedded core. Currently we support the following distinct embedded core test strategies (and mixes of them) in our design system:

- Multiplexing

  All core I/O's are multiplexed to chip I/O's which makes them directly accessible for the automatic test equipment. Existing vectors can be easily reused, but there is a big area overhead.

- Internal Boundary Scan

  Access to core ports is provided via a boundary scan ring allowing full controllability and observability for core ports, but add a huge pattern overhead if used exclusively (cf. Figure 6.8).

Support of these two techniques in the design system is achieved through tools that merge the setup and the production test pattern as well as the mapping of core port names to I/O pad names.

- Full Scan Test

  In this scenario ATPG is run on the complete chip including the core, which fits smoothly into the overall chip manufacturing test methodology mostly used in today's design flows. The drawbacks are that a gate level netlist is required, which is a problem with respect to IP protection, and that all test pattern need to be generated for the chip including the core. Modern design flows should support modularity also in the test area and IP needs to come with pattern of a known test coverage.

- Built In Self Test

  If BIST is built into the core, on chip testing can be done by the chip designer with minimal effort. The problem here is that most cores today do not include BIST, so this remains to be the test strategy of the future.

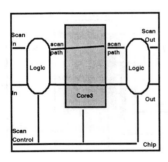 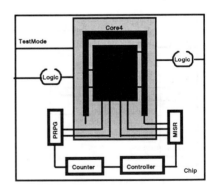

Figure 6.9 Full Scan and built in Self Test

The most effective approach we have seen so far is a combination of these techniques, like:
- Boundary Scan for core I/Os
- Scan for internal core logic
- Multiplexing for memory arrays

6.6 BACKEND VIEWS AND BACKEND DESIGN

The objective of the backend design process is to develop a verified physical layout of the design which then can be used for mask preparation and manufacturing. Usually, a netlist of the design is given to enter this process. Hard IP blocks are instantiated there the same way as standard cells are.

The backend design process consists of the following steps:
1. Floorplanning, Place and Route
2. Extraction of Parasitics
3. Final layout assembly
4. DRC and LVS verification.

Techniques to perform these steps on a design containing hard IP blocks are already available and well known. The approach is similar to integration of fixed layout blocks, like memories, on a chip. Differences are related to the backend views itself and the way they are generated.

At the same time as an already available hard IP block is considered for general reuse in embedded designs, it is recommended to decide which cell library should be provided along with the IP block. Design rules, electrical characteristics, layer lists, etc. have to match to be able to manufacture the IP together with the library elements on the same chip.

Of course, such library compatibility problems could be figured out and fixed still during the backend design process. But it's time consuming and a potential source for errors in the design. Therefore, the compliance of such libraries to modelling standards is an essential success factor for reuse. It is not sufficient, that both libraries just use the same data formats.

To support place and route as well as floorplanning, a special view - an abstract of the IP block - needs to be created. This can be done based on the IP block's GDSII file. The following information is provided in the abstract:
- Pins - define geometric shape and location where a router can connect signals to the IP block.
- Blockages - define geometric shape and location of areas where the router is not allowed to route signals.
- Boundary - defines the outer shape of the IP block.

In addition to the geometric information, data about pin capacities and timings are incorporated to support extraction of parasitics and timing driven Floorplanning, Place and Route. In order to generate an abstract which allows proper integration of the IP block into a chip, some questions need to be answered:

- How to supply the IP with power, ground and clock?
- Is it allowed to use power and ground rings of the IP to supply embedded logic around it?
- Is it allowed to route signal wires over the IP?
- Are there free areas which need to be fully blocked for routing?
- Are there other than minimum width wires required to connect signals?

Depending on such backend integration guidelines, different kinds of abstracts with more or less details can be generated (Figure 6.10). On the left side a full blockage model of an IP block is shown. The entire area is blocked and the router is only allowed to connect wires to the pins. No routing over the block is possible. Except the power and ground pins at the top and bottom, no additional power rails of the block can be shared to supply embedded logic around the IP block. This represents the methodology, how fixed layout blocks had been integrated in the past. To integrate today's hard IP blocks much smoother into a chip, more detailed abstracts can be used.

On the right of Figure 6.10 such a model is shown. It is a "Shrink Wrap" model, based on appearance of it's blockage. The router is allowed to use free routing resources within the block. The entire geometries of the Power and Ground ring's are extracted and available as pin shape. Thus there is a lot of flexibility to supply the IP block and also to connect power stripes of standard cell rows. Due to the more details of this model, it is of larger file size and also routing may take longer than routing using a full blockage model. However, the backend designer has more flexibility and the layout may be more compact using this shrink wrap model.

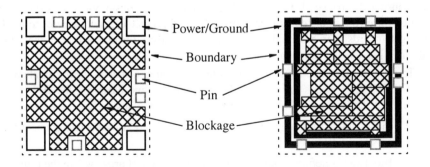

Figure 6.10 Full blockage and shrink wrap abstracts

Today, tools and methodologies to generate abstracts out of layout data are already on the market. Usually, they are provided along with Place and Route tools and are able to generate abstracts directly in the required data format. However, presently those abstract generation tools are mainly targeted for standard cells and less complex macro blocks. In order to generate more details in a large (several hundred pins) abstract manual user interaction is often necessary.

After Placement and Routing of the embedded design is finished, its final layout needs to be assembled. This is done by replacing the abstract view of the IP block and of all other cells in the design by their layout representation. Those layouts are provided as part of the library packages. Commonly GDSII file format is used. The embedded design, created by a Place and Route tool, is available usually in GDSII or in DEF format.

Usually, the layout of large IP blocks, like a microprocessor core, is hierarchical. Sub-modules and cells are put into the layout as instances. But during the chip's frontend design, names of those sub-modules and cells within the IP block are normally not known by the designer. Conflicts will be caused if the sub-modules or cells within the IP block have the same name as modules or cells in the embedded logic part of the design. The tool which performs the final layout assembly (e.g. a layout editor) will find 2 layout sources for such instances - one in the GDSII file of the IP block, and another one in the GDSII file of the cell library (or in the GDSII file of the embedded design itself, if sub-module names are duplicated). Since it is not possible to have different sub-modules or cells with the same name in the final GDSII file, those names need to be changed to be unique. This problem can be prevented by adding a prefix (e.g. the IP block name) to all sub-module and cell names of the IP block. The resulting naming style of IP block's internal modules and cells should be provided as part of the module documentation to the designer.

Additionally to preparation of IP block's GDSII file, it's transistor level netlist needs to be supplied for LVS verification. Here the same modification to cell and module names should be done to avoid name clashes when the IP block netlist along with transistor netlists for embedded logic cells, and along with a top level netlist of the design is read into the LVS verification tool.

6.7 IP REPOSITORY

Having a central repository to distribute IP is an important vehicle to steer reuse. The Motorola Module Board and the underlying Minerva system were created two years ago already to facilitate an intra-company repository. Its main focus was on soft IP and related topics (documentation and coding standards, compliance checking tools, peer reviews, etc.). Recently we have enlarged this towards hard IP and are in the process to add software IP and other intellectual property (e.g.: testbenches). With this new focus we also changed the name from Module Board to IP Repository.

The architecture of the IP Repository is composed of four major building blocks: The graphical user interface as a presentation layer for the different types of users, the IP meta data server to store any meta data about IP, multiple distributed IP vault servers to store the IP data and the vertical IP data distribution to deliver IP.

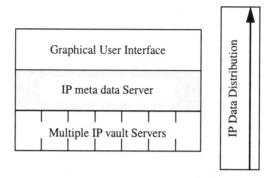

Figure 6.11 IP repository architecture

The users of the system fall into two general categories: IP provider and IP consumer. The IP provider have to have the capabilities with the user interface to maintain their IP and its IP meta data, while the IP consumer needs powerful search and comparison functionality to find efficiently necessary IP. The following is a very condensed description of the main functionality necessary to build an IP Repository system:

1. **User Interfaces**
   A web based graphical user interface should provide capabilities to search and browse through the available IP. Different search methods could be combined to find the right IP. Keywords search, category based search (taxonomy) or filter capabilities through IP meta data pre-selection should lead to the requested IP. Similar IP should be possibly compared against each other, to ease the decision for the IP consumer. The discovered IP data should be can be down-loaded directly to the IP consumer.
   Usage statistics and reports are very essential as well, for example to convince IP consumers of the quality of an IP. The confidence of an IP will grow, if it has been used already successful on silicon, and this is reported for the IP. An administrative interface is vital, to maintain all IP meta data as well as setup the access control information for an IP block.
2. **IP Meta Data Database**
   Each IP might have any additional information about it, to further describe its behaviour. This additional information is called the IP meta data, which would be stored for example into a central relational or object database. It

will be queried by the user interface to find the requested IP. To be extendable, the database design should provide a mechanism to easily enlarge the possible IP meta data definitions as necessary without switching of the system, or even the need to rebuild the system. It is essential for the usage, that the design is as data type independent as possible, so it could support any kind of IP in the same IP Repository, like design IP or software IP.

3. **Data Storage and Configuration Management**
   IP data should be stored within a configuration management system to ensure the traceability of different versions of IP within the system. The IP vault servers could be distributed throughout the Intranet, so that the IP providers can control and maintain their own pool of IP

4. **Security and Access Control**
   Authentication, normally done via an account and password, ensures the identity of the user. Authorization, checks the rights a user have to access IP. An authentication is a prerequisite for the authorization. Despite internal access restrictions for an IP, an international company have to additionally comply with the export control restrictions for the different countries and have to enforce export control rules even for their internal use, e.g. encryption IP.

5. **Change Management System**
   Often, IP consumer will have change requests for IP, while integrating it into their system. The system shall provide a change management system to give the user this feedback mechanism to the IP provider, but also to any other interested IP consumer of this IP via customized notification agents.

Some non-functional requirements are quite essential as well for a reliable system. On-line backup of the data is vital for 24x7 hours availability, while even fault tolerant services have to be considered in an international environment. Using english as the only language on the user interface might be not convenient enough for the user to use the IP Repository at all, so preferably it should be a user preference to setup. The data and time format should be changed accordingly. Proper documentation for all the different user interfaces needs to be available, as well as an on-line training to encourage the use of the system.

There are already systems available like the catalogue server from the RAPID organisation or from "Design And Reuse" (http://www.design-reuse.com), but even those fulfil just partial the high level requirements described above.

6.8 NEXT STEPS

Future activities will be to add new cores (STAR12 and DSPs), further refine the methodology and get in line with industry standards like those defined by VSIA. We have already started to fully automate the characterization and view generation processes, and aggressively drive the adoption of reuse tools, methods and standards throughout Motorola and our customers. A special focus will be to further populate the repository and accelerate the reuse of Motorola IP inside, and eventually also outside of Motorola.

Wolfgang Eisenmann received his Diploma and Ph.D. degree in Electrical Engineering from the Technical University of Munich, Germany in 1990 and 1996 respectively. From 1990 on he is working with Motorola GmbH in Munich, Germany. His experience includes ASIC design, delay and power modelling, timing/reliability simulation and IP reuse methodology. Currently he manages the IPworks-Europe group in Munich, which is part of the SoCDT organization within Motorola.

ColdFire and M·Core are trademarks or registered trademarks of Motorola, Inc. The PowerPC name and PowerPC logotype are trademarks or registered trademarks of International Business Machines Corporation, used under license therefrom

7 A REUSE LIBRARY APPROACH IN ENGINEERING CONTEXT

S. Müller

Corporate Research and Development Systems 2
Advanced Development Multimedia Systems
Robert Bosch GmbH, Hildesheim, Germany

7.1 MOTIVATION

With the increasing number of designs developed in various projects, the need for a library comes up that archives designs and modules in a convenient way for later reuse. The intention is to provide all necessary information that describes the module to allow an early reuse decision against re-building from scratch and most important to reduce design time by reusing.

But, most of the designs and models archived cannot be reused without modification, they have to be adopted to the new application. The decision for reuse is a process that requires useful information about the concerned module and it requires to consider the modification level. But even the reuse of an existing model as a template or basis for a new development is enough motivation to set up a library and to assemble reusable elements.

High efforts for module validation and verification can be minimised when a new development is based on the experience of already finished designs with proven functionality, e.g. design safety is increased.

Beside these project-management items, the existence of a reuse library represents know-how and knowledge and demonstrates competence. Therefore, advertisement in the circle of library users is essential to increase the acceptance of the library and encourage designers to use elements and submit their new developments. Efforts to keep quality level and maintenance high are to be considered.

Usually, not all modules are designed for reuse or are not even applicable in other designs. The intention of this contribution is not to characterise the different stages of reusability, but to show the way to develop and implement a practicable reuse system in the design environment of a system and chip designer.

7.2 PROJECT DESCRIPTION AND OBJECTIVES

The need to reuse already developed modules with proven and verified functionality is caused by the growing complexities of today's systems in order to fulfil given time-to-market requirements. In our department, reuse mainly took place within project teams where designers work together. Here, communication between the designers was practised because of the common project goal. But, reuse efforts between different projects were low even if the designers involved worked nearby.

Furthermore, it was discovered that possibilities are missing to broadcast and archive design ideas and developments and if designers leave a project team, know-how about design ideas would have been lost as well.

Regarding the design flow in a project, modules are described in models of different abstraction levels at different project stages. These models describe the same functionality of one module. This is only indicated by archiving these models in the same project directory with more or less useful names and even sometimes together with other modules of the concerning project.

Furthermore, the entire design of a project is often too complex and too specialised to be completely reused. Convenient module sizes as part of entire design in our projects had not been identified and explicitly declared for reuse at this time.

As a consequence of the results mentioned, it was decided to emphasize on the exchange of designs and design parts between projects and to ease communication by help of a tool: a reuse library. Style guides and design rules had to be written that describe how modules are to be developed and prepared when the intention is to check them in. Moreover, style guides and design rules have the purpose to increase readability, transparency of the developments, and quality. At least the availability to other designers should be enhanced. Before a methodology had been implemented, designers were involved in the developing process of the library concept.

7.3 REUSE METHODOLOGY

Designers had been invited to contribute to a basic concept for a reuse library. Finally it had been decided that the approach should be web-based and each library element should be represented by one coversheet that offers a short description of the design functionality and important design characteristics at a glance. Sources and more detailed documentation can be down-loaded on demand.

The library is defined as a so-called copy system (in contrast to a link system) where data is copied from the project directories into the library data directory that is physically located at a different place in the file system. While a link system would provide links to desired data or even the file system path where the data is located.

With the help of the meta-plan technique, six conceptual parts were identified that had to be prepared until the entire concept for the reuse library could be defined: Data structure of required information about a library element, submission flow when applying a module to the library, design rules and style guides for the different expected data types of a library element, access regulations that explain who may use the library and who may have access to specific data in the reuse library and a man machine interface.

In the following, results of the conceptional work in the work groups are presented.

7.3.1 Concept Partitioning

The library project had been analysed with respect to obligatory concept parts, and with respect to the required communication effort between the design groups. The result is shown in the schematic in Figure 7.2. Each circle describes a work group that had prepared a part of the concept of the library, before the reuse library prototype had been implemented. The direction of an arrow indicates the information flow during development of the reuse library concept, e.g. a work group had to wait for results of an other work group.

Two work groups are independent from results of other work groups: Firstly, the "Data Structure" group explains which data is necessary for a module to become an element in the reuse library. The "Access Regulations" group that determines possible users of the library and who may access which parts of the reuse library. Secondly, additional results come from the "Submission Flow" group and the "Man Machine Interface" group. The "Submission Flow" group offers all mechanism to allow easy submission of a new module to the library. The "Man Machine Interface" group specifies procedures (script system) that are necessary to handle data between user and library when submitting a new module to the library or checking / searching the library for available modules and down-load sources.

82   REUSE TECHNIQUES FOR VLSI DESIGN

The "Rules and Style" group defines the guidelines of the module development of the reuse library. Style and contents of the coversheet has been defined by this group.

The "Presentation" group collects and displays information to guide the user through the library. The look and feel of the tool is important for the acceptance of the library by user.

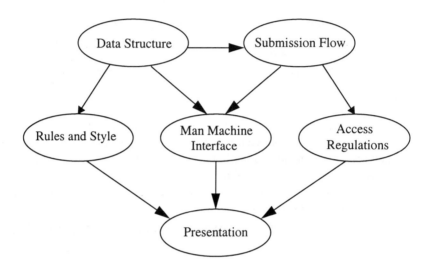

Figure 7.1 Work groups and dependencies on results

7.3.2 Data Structure

The following data types can be stored in the library: C, VHDL, C++, MATLAB, tests, scripts, tools and documentation and "other". The last category is open to be named by the submitting designer when using a special tool with a special design data format.

For the prototype library, an underlying database system is not considered, but a simple directory structure is used. The expected number of elements in the library can be easily managed by this structure.

Each directory contains all data belonging to one module that is represented by one coversheet at the presentation layer. Furthermore, subdirectories contain module data at the special description levels or they contain additional data like documentation, tests, and simulation and synthesis scripts.

For example, a library element for a VHDL module may offer the VHDL sources of the module at behavioural and/or register-transfer level. Furthermore, it contains a testbench, scripts to start a VHDL simulation and logic synthesis, stimuli and reference data to perform the verification. The reference model may be written in C language and it may run under MATLAB, and therefore, the MATLAB simulation configuration is also available. In conclusion, all data is stored in the reuse library that is required to rebuild the design of the module at each description level and to re-run simulation and/or logic synthesis. Data that belongs to special VHDL entry or design tools can be stored in an extra directory.

An other approach was to insert the modules in fixed categories (taxonomy). But this approach was not implemented since the reuse library contains a powerful search engine. A keyword catalogue is planned.

7.3.3 Access Regulations

For security reasons, it is important to keep "intellectual property (IP)". A dedicated web server has been installed for the reuse library to protect the IP. It has been decided to take the SUN web-server delivered with Solaris 2.6 This software is based on a three level security system.

First of all, an administrator access exists that allows to change the structure of the library, e.g. initiate new elements. A limited number of persons have administrator access to the library and they can handle submission requests from users. Passwords are mandatory for full access and to obtain data and sources of reuse modules in the reuse library.

The guest account is the lowest access level and it has a read-only permission for coversheets and global information like design rules and style guides. The sources are not accessible for guest accounts.

7.3.4 Rules and Style

Design rules and style guides are prepared for C, MATLAB, VHDL and the coversheet to provide design transparency and a certain level of quality for modules. Global design rules had been defined that are obligatory for all projects. Besides defining the available constructs, it supports the designer to easily navigate through a design by useful chosen conventions and names. Each project must have special design rules since, e.g., memories have to be always instantiated in different ways for different projects. Project specific design rules are also bound to the reuse library.

7.3.5 Quality Aspects

The work groups "Submission Flow" and "Style and Guide" are mainly responsible to consider quality aspects for the reuse library. A fully automated element insertion system for the library is not implemented. Due to that, communication between

84 REUSE TECHNIQUES FOR VLSI DESIGN

the administrator, who is an experienced designer, and the designer has to take place to guarantee quality, plausibility and completion of the library elements, e.g. because low quality decreases motivation and acceptance of a library. To guarantee quality both design rules and style guides have to be periodically reviewed.

The coversheet presents the status of a module. Additional information is provided for the grade of verification of the module. Further, deviation of the functionality of modules may appear after the component is stored into the library. For this case, a distributed defect tracking system used to archive bugs and to reflect the current state, became a standard tool in the design flow.

7.4 MODULE ADMINISTRATION

Module administration includes the module submission and the library maintenance. The main strategy for the designer is to keep these efforts as low as possible.

The designer completes a predefined coversheet. The administrator displays the incoming temporary coversheets, implements a new library element and updates the library element list at his administrator page. This page holds a variety of cgi-scripts and allows an automatically handled module submission from the web-browser window.

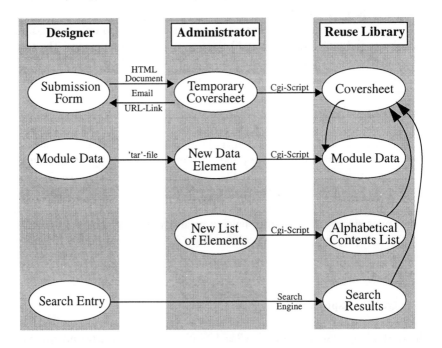

Figure 7.2 Library submission flow

A REUSE LIBRARY APPROACH IN ENGINEERING CONTEXT 85

Figure 7.2 shows the submission flow. A designer adds design characteristics into the pre-defined submission form to describe his module. He sends it to the administrator by simply pushing the email button at the bottom of the document. He may attach the sources of the modules to this email or send it later.

The administrator receives the email and rebuilds the coversheet with a cgi-script. The incoming submission form can be displayed with an Internet browser, again, and becomes a "temporary coversheet". This can be edited, so that the administrator inserts questions and comments when he does not agree on the contents of the "temporary coversheet". Then, he sends it back to the submitting designer.

The submitting designer receives an email containing an URL-link to the modified "temporary coversheet". Thereafter, he can check the "temporary coversheet" and he has the possibility to change the contents if changes are required. The "temporary coversheet" appears to the submitting designer like the submission form. After the missing data have been inserted, the designer sends it back to the administrator.

The document will be circulated between administrator and designer until a final state has been reached. The "temporary coversheet" will be stored into the reuse library to represent the module, and therefore, it becomes the final coversheet.

Cgi-scripts that are applied from the page of the administrator fix the final coversheet, import it into the reuse library, and set up the data structure for the new library element.

The library contents is automatically updated by a further cgi-script. Figure 7.3 shows an example of the entry in the library for two letters "S" and "T". The associated modules of the library are presented. Since there is no library element available starting with the letter "t", the comment "empty" appears.

| S | • ScaleImage | Performs scaling (interpolation) of an 420 YUV input image. Scaling is done by bilinear interpolation. C-function with Matlab test environment. Integer arithmetic!! |
| | • SharpenImage | Performs a sharpening operation to the luminance component of an input image by applying a 2 dimensional filter mask. C-function with Matlab test environment. Integer arithmetic!! |

↑ Seitenanfang

| T | | empty |

↑ Seitenanfang

Figure 7.3 Example of the library contents list

86   REUSE TECHNIQUES FOR VLSI DESIGN

The effort for administrator and designer for adding a new model into the database of the library is considered as low. Nasty overhead is handled by cgi-script. Both persons can focus upon the re-useable implementation of library elements and completion of available information and tests. The communication between administrator and designer is obligatory to enhance quality and it fulfils the strategy of the re-use concept.

7.5 PRESENTATION AND ACCESS TO MODULES

The designer who is going to access the reuse library can obtain the data of a module by using a search engine or browsing the alphabetical module list. Both the alphabetical list of available modules and the result of the search link to coversheets of the modules. Buttons are located at the bottom of each coversheet to down-load the sources of a module.

```
module name, project name
author, phone, department, email address, date
short description of the functionality
existing file types for the module (C, VHDL,...)
if module is verified, comment
known faults, comment
estimated grade of re-usability
if design consists of RAMs, comment
hierarchical design structure, comment
if testbench exists, comment
if simulation scripts exit, comment
if synthesis scripts exist, comment
if stimuli exist, comment
used (design) libraries, comment
if additional components needed for verification, comment
existing abstraction level for functionality (models)
clock frequency
used tools
silicon technology, gate count
already applied in which design
```

Table 7.1 Comment text

In Figure 7.2, arrows in the column "Reuse Library" show that the coversheets may be accessed via links in the "Alphabetical Contents List" of library elements, see Figure 7.4, or links presented in the "search results" of the search engine. Access to

A REUSE LIBRARY APPROACH IN ENGINEERING CONTEXT 87

the module data in the reuse library occurs via coversheets, again in Figure 7.2, an arrow is pointing from the "Coversheet" to the "Module Data". A coversheet for a library element contains the following information where "comment" shows that arbitrary text can be included (cf. Table 7.1):

Since it is possible to insert different library elements (VHDL design, C-routine, or even a plain documentation...) some lines of the coversheet may remain open. Administrator and designer will determine which information is required for the coversheet to insert a certain library element.

Figure 7.4 shows the home page of the reuse library "Multimedia Library" where also additional services are offered to increase its attraction. The links are mostly named in German, the translation follows figure 4:

Figure 7.4 Entry page of the reuse library

The ability to allow access to/from other libraries within Robert Bosch GmbH is planned and first steps are done by exchanging the element lists of the other reuse libraries.

7.6 MEASUREMENT

Measurement is important for designers who want to estimate their effort to rebuild the database of the module and re-run simulation with the purpose to understand the module functionality in greater detail.

An easy way to define a measurement for re-usability is to estimate the required day count that is necessary to re-build the database of a model and to re-run simulations. The inverse of the day count is an expressive number. For example: an estimated day count of 5 comes to 20% reusability, one day means 100% reusability. The required time to set up again the design is important if reuse should be considered.

7.7 SUMMARY

The development of a reuse library has been presented. In this study, designers had been early involved in the concept and implementation phase of the reuse library to ensure its later success and acceptance.

Library elements can be routines of high-level programming languages, designs that are modelled by hardware description languages, a documentation only or interconnection of different modules that can be stored in the reuse library as well. More complex library elements may contain models of different descriptions of its functionality at different abstraction levels. One library element may include the VHDL source, its reference model in C and the obligatory data to re-run simulations at each abstraction level or logic synthesis.

Additionally, documentation is required to explain the module, showing the different existing descriptions of the functionality, design characteristics, and global information about the module. A document "coversheet" has been introduced that presents a library element. Designers may access the coversheet of a library element as a result of a search engine or they can chose the coversheet out of an alphabetical list of all library elements.

Since quality of the modules in the reuse library is important, especial emphasis was put on the communication between the submitting designer and administrator of the library. A web-based approach has been chosen and the submission and the access to library elements can be performed by a web browser and email.

Using email and web browser, the coversheet can be exchanged between designer and administrator until both agree on a final version.

Furthermore, design rules and style guides are prepared to enhance the quality of the library elements and support re-usability for new designs, e.g., project start of a new system or chip design.

7.8 CONCLUSION

The best library is useless if nobody employs models: the library is not "alive" and stops representing competence and knowledge. The reuse library must be a part of the working environment of each designer to fulfil the high expectations. Quality is one important point that may attract designer to accept the reuse library as a common tool in his environment. But how get designer used to it?

We identified the development of testbenches as a field in the daily work of a designer where reuse is exceptionally often needed. Our testbench approach is the use of modular testbenches based on graphical entry tools. Testbench modules like clock and reset generator, file I/O, checkers, interface simulators, etc. can be submitted to the reuse library like other design parts or modules. Each designer has to set up testbenches and he can find useful modules and templates in the library. As a result, the library grows and designers get used to work with the library.

7.9 STATUS OF THE WORK

After setting up the prototype of the reuse library, the submission of modules started. At the beginning simple modules have been included that consists only of one model, e.g., the description of its functionality. This was performed to test and evaluate the submission flow and the implemented mechanism that had been easily handled as we expected.

During start up, designers checked new modules to get an idea of the technical solution contained in the library. Feedback had been received, because some modules names and keywords have not been carefully chosen and provoked confusion about the functionality. As a consequence, useful names and keywords had been chosen to support an early understanding of the technical intention of the module. This accelerates the reuse decision for or against the module. Due to intensive communication between designers: some designers became known as experts because they inserted their modules into the reuse library.

7.10 OUTLOOK

In the future, more effort will be spent on the selection of keywords and to develop keyword catalogue that is available to designer. A keyword catalogue is obligatory if modules are going to be exchanged between libraries.

Several departments at Robert Bosch GmbH already run reuse libraries. The idea is to make the modules in the different libraries accessible to other designers. This will force conceptual work on interfaces between reuse libraries or the establishment of one powerful reuse library that contains all available modules of all these libraries. First steps are performed to exchange data between these libraries.

 **Steffen Müller** received his degree from Kaiserslautern University, Germany and continued his studies at the Michigan State University as a scholar in 1992. From 1993 to 1997, he worked at MAZ Hamburg GmbH where he was involved in the development of dedicated hardware for digital image processing. He is currently with Robert Bosch GmbH, Advanced Development Multimedia, where he is responsible for design methodology. His VLSI development activity is in the area of digital audio broadcasting (DAB).

8 ASPECTS OF REUSE IN THE DESIGN OF MIXED-SIGNAL SYSTEMS

F. Heuschen, Ch. Grimm and K. Waldschmidt

Johann Wolfgang Goethe-University
Department Technical Computer Science
Frankfurt, Germany

8.1 ABSTRACT

A well structured top-down design flow is based on different levels of abstraction. Each level operates on data types that reflect the level of abstraction. Therefore the design flow demands a well structured data organization. Not only the design flow but the design itself is improved by *reuse* of previously designed or verified data. Storage and organization has to be taken into consideration as well as the concrete interface between data and objects that use it.

This paper describes where and how a top-down design flow for mixed signal systems can be efficiently supported by outsourcing existing data into a relational database system, which is then supplemented by new data as it is created.

8.2 INTRODUCTION

The exploitation of *re*-use of any applicable data to further reduce design cycle times, is of wide interest in research and development, but it is still rather uncommon for analog or mixed signal systems.

The design of mixed signal systems in general is not yet so well structured as the design of pure digital systems. The reason is that the design of the analog part of a hybrid system is still *bottom-up* and therefore is more time consuming and requires much expert-knowledge and experience. It is desirable to create mixed signal systems with a common design methodology that cares for common constraints and resources across the boundaries of partitions. Such a design methodology is described in detail in [Grim98b].

A unified design methodology that allows abstract specification of mixed signal systems offers the opportunity to also integrate reuse aspects for the analog part of hybrid systems. The reuse methods described in this paper are applicable both for system parts that are specified as differential equations as well as system parts that work as processes.

A design tool based on the top-down design flow and on mixed signal reuse databases would finally support an experienced designer in finding the "*right*" partitioning of a usually over-specified hybrid system for the intended system functionality. Over-specified because specification of an intended system behaviour often implicitly contains a kind of structure. This is explained in detail in the following section.

8.3 TOP-DOWN DESIGN FLOW

To structure the design of mixed signal systems we define three different levels of abstraction:
- System Level: Specification of functional and non functional properties
- Architectural Level: Partitioning, introduction of structure
- Circuit Level: Concrete representation of technology

We especially concentrate on the difference between "behaviour" and "structure" to distinguish this approach from a pure structural refinement [Grim98b]. At the **system level**, the intended system's behaviour is specified together with non-functional constraints (cf. Figure 8.1).

Because of the mixed properties, mixed signal systems have to be specified in a heterogeneous way, using both continuous-time differential equations and discrete-time processes, separated by the structure of a block diagram. Unfortunately the block diagram structure suggests interpretation as the intended structure of implementation, even if the intended behaviour can be represented in a different or usually more optimized structure.

Therefore a step of repartitioning is done to find alternative structures and the one that best suits the specified behaviour with an efficient usage of the given

ASPECTS OF REUSE IN THE DESIGN OF MIXED-SIGNAL SYSTEMS 93

resources [Grim98a]. Note that this step means more than a structural refinement: the resulting block diagram is repartitioned to new blocks that can consist of parts from several different blocks from the initial diagram. Another kind of repartitioning could possibly result in the complete and unmodified transposition of a certain block's time model from continuous-time differential equations to a discrete-time process and vice versa. Repartitioning is based on *knowledge* about existing circuits and physical effects as well as on *estimation functions* that quickly approximate the resource requirements for different kinds of implementation.

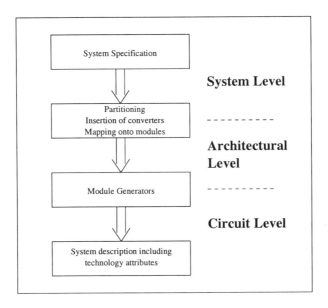

Figure 8.1 Design flow for mixed signal systems

At the **architectural level** we find now a block diagram that is optimized for the concrete implementation. The blocks represent modules that can be designed in a less creative and more automated way. Converters between different types of signals are inserted into the block diagram if needed and all together the blocks are mapped onto *module generators* where the design is interactively and automatically further refined. Module generators are a set of small software modules that automatically transfer the intended functionality to the physical world of electronic circuits and automata. In the back end of this step the system reaches the **circuit level**, where it is represented by its concrete components such as operational amplifiers, resistors, capacitors, gates and registers.

On its way through the levels of abstraction, the designed system is processed as a homogeneous hybrid dataflow graph that is basically an extension to the dataflow principle and introduces an activation rule for *each* single node [Grim96].

A second but very useful property of index hybrid dataflow graphs is, that groups of nodes can be assigned to modules that can be designed independent from each other. Design and depth of refinement of modules can be done in an arbitrary order. Construction results are added to the concerned nodes as *attributes*. So the hybrid dataflow graph represents the system at any level of the design flow and at any state of refinement.

8.4 DATABASES AND REUSE

We can identify at least four points in the design flow described above, where the reuse of data, previous construction results and *intellectual property* (IP) is reasonable:
- at the *system/architectural level*: mapping onto modules
- at the *architectural level*: module generators use IPs
- at the *architectural/circuit* level: module generators use technology tables
- at the *circuit level*: setting up the attributes with generic module descriptions

This is visualized in Figure 8.2 (the terminology is explained in detail below).

The reuse of previous construction results and *intellectual property* (IP) demands also, that simulation and verification are kept in mind. This requires at least one further collection of data, storing *behavioural models* to speed up the simulation process.

8.4.1 Mapping onto Modules

In the *module mapping* step of the design-flow the partitioned hybrid dataflow graph has to be analyzed to identify functional related blocks that can be processed by different existing module generators.

A mapping algorithm identifies all sub-graphs that match existing types of modules and afterwards chooses the modules that integrate functionality in an optimized form. Unfortunately this contains a set covery problem as well as a graph isomorphism problem that are both known as *NP*-complete, but heuristic based algorithms can figure the mapping out in an acceptable runtime.

ASPECTS OF REUSE IN THE DESIGN OF MIXED-SIGNAL SYSTEMS

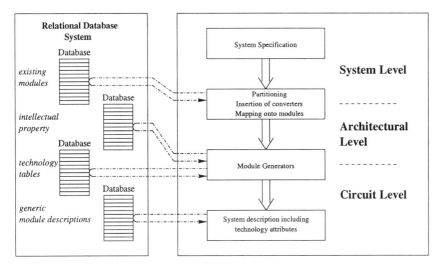

Figure 8.2 Design flow extended by a reuse database

The mapping algorithm needs exact information about the module types and their generators to perform this identification. Information about existing generic modules and their attributes should be known. The attributes of a given module generator should consist of:

- a characteristic section of a hybrid data-flow graph's node that identifies the module type
- name of the module type
- invocation of the module generator
- arguments of the module generator invocation

One way to accumulate this information is to set up a list in the program code that is iteratively searched by the module mapper.

A better way is to incorporate the information in a data table that is managed by a *relational database system*. The module mapper analyses the hybrid dataflow graph, extracts characteristic attributes out of the nodes and performs a SQL query to find the module generator that suits to that kind of node. Modules with more complex functionality can in return shift the bounds of the identified blocks.

The benefits of the database solution are evident:

- The module mapping system remains compact and concentrates on the essential problem, the pattern matching.
- The outsourcing of data to a secondary system or even to a secondary server computer might improve the performance.

- Data are expandible. Additional module generators can be implemented at a later time and then simply included in the mapping process.
- Data are reconfigurable without changing the mapping procedure.

8.4.2 Intellectual Property

An important aspect of reuse is the consideration of using intellectual properties, or IPs. *Design by reuse* and *design for reuse*, the construction and use of IP-libraries, is of wide research interest. A formalized method of building IP-libraries for digital design can be found in [Koeg97], [Koeg98].

Nevertheless, the building of IP libraries for analog or mixed signal design is rather uncommon. In our view an index IP reuse should base on three different types of IPs on different levels of abstraction:

- Previously designed modules
- Pre-designed parts
- Parametrizable templates of standard cells

The module-mapping system identifies parts of the hybrid data-flow-graph for the assignment to a module class. Afterwards an IP library should be searched to identify a suitable *previously designed module* that is directly applicable or can be used with slight modifications. Subsystems with a certain bandwidth in their functional constraints, for example A/D-converters, filters or integrators, can be handled this way. These kind of IPs are stored in a *relational database*-table, where they can be queried for identification.

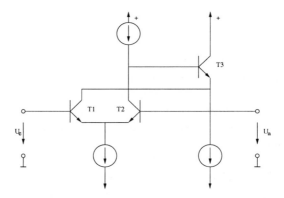

Figure 8.3 Internals of an operational amplifier

ASPECTS OF REUSE IN THE DESIGN OF MIXED-SIGNAL SYSTEMS 97

Pre-designed parts are the second kind of IP on which calculation templates are based. Examples include pre-designed operational amplifiers (Figure 8.3) and gates.

Their values and properties can be found in *technology tables*, where the module generators can retrieve them for calculation of resource requirements.

The third kind of IP, the *parametrizable standard cells* are stored in the appropriate module generators.

Module generators contain templates of circuits that are based on the usage of simple electronic or predesigned parts with a slightly more complex body.

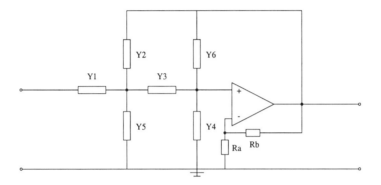

Figure 8.4 Template for Sallen-Key filter

Figure 8.4 shows a template for second order Sallen-Key filter with an abstract representation in form of admittances rather than actual components. This permits the usage of the same template for a lowpass, bandpass and highpass filter by interpreting the admittances Y_1-Y_6 according to the intended functionality [Morg97]. This template is contained in the *analog filter* module generator.

8.4.3 Technology Tables

If no suitable IP was found in the library of previously designed modules, the intended functionality has to be realized by the appropriate module generator. The functional demands lead to refinement down to net-lists or register-transfer representations. To determine the resource requirements of a given module the non-functional properties are calculated or estimated from the amount of operational-amplifiers and the values of resistors and capacitors in the analog domain or the amount of gates and literals in the digital domain.

98   REUSE TECHNIQUES FOR VLSI DESIGN

The relation of an object to its non-functional properties is usually set in *technology tables*. These are directly transferred to data-tables that again are supervised by a *relational database* system.

Table 8.1 shows a data definition that allows a technology independent design. In Table 8.2 an example from analog domain is given, showing how the technology table is filled:

Field Name	Data Type	Field Size (Byte)	Description
technologytype	varchar	20	name of technology
technologysize	float	8	base size in m
supply_voltage	float	8	supply voltage
element	varchar	20	electronic element
unit	varchar	4	unit
unit_exp	int	4	value of exponent
area_per_unit	float	8	area in m / (unit*E unit_exp)
power_diss	float	8	power dissipation in Watt
power_diss_exp	int	4	value of exponent

Table 8.1 Technology table field definition

technologytype	...	BiCMOS	BiCMOS	...
technologysize	...	1.2		...
supply_voltage	...			...
element	...	Capacitor	Resistor	...
unit	...	F	Ω	...
unit_exp	...	-15	3	...
area_per_unit	...	1	2	...
power_diss	...			...
power_diss_exp	...			...

Table 8.2 Technology table example

A Capacitor in 1.2 m BiCMOS technology occupies 1 m per fF (*femto Farad*). A resistor of same technology uses 2 m per $k\Omega$. Power consumption is signal-voltage and frequency dependent and has to be calculated separately.

Technology dependent parameters are retrieved by SQL-query and multiplied with the actual values calculated by the generator to get non-functional properties like area, power consumption and delay.

This outsourcing of technology data has the following advantages:

- separation of code and data
- data, that is used by a number of different module generators exists at a single point of administration
- the technological progress is taken into consideration by simple extension or replacement of the data-tables

8.4.4 Generic Module Description

In the backend of our design flow the system is represented by a fully attributed *hybrid data flow graph*. It is processed as a ASCII-text file. Each module generator has to produce his specific part of this attributed graph, and in the end when the complete system is constructed all parts of the graph are concatenated to the system graph. It is reasonable to define generic module-dependent graphs that are rather parametrized than created by the module generators.

```
( view dess
  ( graph integ
    ( node list
      ( node a_integ
        (type INTDT)
        ( in-port in_signal a )
        ( out-port out_signal x )
        ( AS-POWER            )
        ( AS-AREA             )
        ( AS-TIME             )
        ( AS-NOISE            )
        ( AS-R                )
        ( AS-C                )
      )
    )
    ( edge-list
      .
      .
      .
    )
  )
)
```

Figure 8.5 Module description template excerpt

Figure 8.5 shows a part of such a generic graph for an analog integrator. The parametrizable values are filled in the (**AS-xxx**) lines, that contain the node's attributes. This generic graphs again can be stored in a table of the *relational database system*. As a module generator has accomplished its task, they are retrieved out of the database, parametrized with the calculated values and then disposed to the supervising construction tool.

8.4.5 Behavioural Models

A top-down design methodology working with reuse needs consideration of behavioural models to speed up the design and especially the simulation and verification process. In particular the design of analog subsystems of a mixed signal system is based on the usage of SPICE model.

According to the three different types of IPs on different levels of abstraction described in Section 8.4.2 a reuse database system should contain stored models for:

- Previously designed modules
- Pre-designed parts

For complete *previously designed modules* as well as for *pre-designed parts* like different types of operational amplifiers behavioural models should be made and stored in the reuse-database system. In future designs they help to build and to *verify* new systems that are based on reusable parts and subsystems.

In analogy to the parametrizable templates contained within the module generators there could be parametrizable SPICE-circuit descriptions for each analog module generator.

```
highpass 50K

.options list node post
.op
.ac DEC 10 5k 500k
.include 'atl081c.inc'
.print ac v(3) v(5)
VCC VCC GND +5V
VEE VEE GND -5V

v1 5 0 ac 0.6v
c1 5 1 24p
r1 1 GND 132k
r2 2 3 71k
r3 2 GND 71k
x1 2 1 3 VCC VEE ATL081C

.end
```

Figure 8.6 SPICE input file example

After the generation process is accomplished the calculated values of resistors and capacitors for example can be filled in such a SPICE-circuit description and the module is ready for simulation. This parametrizable SPICE input files can be stored in a table of the *relational database system* according to the *generic module descriptions* described in the previous section. Figure 8.6 shows a SPICE input file example of a 1st-order Highpass with 50kHz cutoff-frequency. In this example we can see the slots for electronic elements as capacitors and resistors as well as the inclusion of an operational amplifiers behavioural model.

8.5 SUMMARY

At the moment, reuse is only common in the design of digital systems. We have presented a top-down design flow for mixed analog/digital systems as a starting point for design space exploration, design automation and reuse also in the analog and mixed signal domain. Then, we made basic considerations about enhancement of this top-down design flow by including a reuse database concept as an integral part of design methodology. We have discussed several points of the design flow, where reusable data is needed and can be stored in tables of a relational database system. This allows us to define the structure and organization of a reuse database. According to the different types of data held in data tables the definition of appropriate interfaces from different design tools to a relational database system should be possible also.

In our future work, we will gather data either with the help of cooperations with other institutes and industrial companies, or with design tools developed or in development at our department. Then based on an sufficient amount of data we will implement such a reuse database system. First for more interactive use by an experienced designer. Later, we will add interfaces to the module generators to integrate a design tool for mixed signal systems that follows the described methodology.

Frank Heuschen is a research assistant at the Technical Computer Science Department at the Johann Wolfgang Goethe-University of Frankfurt. His main field of research is design methodology and automation for mixed analog/digital circuits, especially partitioning into analog/digital domains. He received his Dipl. Ing. degree in Electrical Engineering from the University of Dortmund in 1993. From 1993 to 1996 he was project engineer at AEG in the systems and automation department for petrol and gas pipelines. Since 1997 he is in the scientific staff with professor Waldschmidt.

Christoph Grimm is a research assistant at the Technical Computer Science Department at the Johann Wolfgang Goethe-University of Frankfurt. His main field of research is design methodology and automation for mixed analog/digital circuits. He studied Electrical Engineering at the TH Darmstadt and the Ecole Centrale de Lyon and received his Dipl. Ing. degree from the TH Darmstadtin 1994. Since then, he is in the scientific staff with professor Waldschmidt.

Klaus Waldschmidt heads the Technical Computer Science Department at the Johann Wolfgang Goethe-University of Frankfurt. His research and teaching interests include computer architecture, especially associative memories and processors, and CAD, especially of mixed analog/digital circuits. He received the Dipl. Ing. and Dr. Ing. degrees in Electrical Engineering from the TU Berlin. He is a member of the Society for Information Technologies (ITG), GI and Euromicro.

9 DESIGN REUSE EXPERIMENT FOR ANALOG MODULES "DREAM"

V. Meyer zu Bexten and A. Stürmer

TEMIC Semiconductors
CAD Services
Ulm, Germany

9.1 INTRODUCTION

The reuse of design knowledge is one of the methods that can be utilized to improve the efficiency of circuit design. This improvement is essential to deal with continuously shorter product cycles.

Especially in the digital domain, a large market for "intellectual property" is already established, allowing companies to sell and buy circuit blocks on different abstraction levels [Behn98, Oehl98]. Also within single companies, reuse methodologies can reduce time-to-market and overall design effort [Muel98].

This contribution discusses different issues of reuse in the analog and mixed signal domain which is characterized by much more complex interfaces and specifications. After an overview of the specific requirements, the implementation of our Design Reuse Experiment for Analog Modules (DREAM) [Stue98, Meye98] is outlined, corresponding experience is discussed, and an outlook on future work is given.

9.2 REQUIREMENTS FOR REUSE OF ANALOG BLOCKS

The main focus of this work is the support of reusability for analog circuit blocks and design know-how. At least if the blocks have to be integrated on chip level, analog circuits depend very strongly on the surrounding circuitry. Main parameters like signal levels and supply voltage can much less be standardized than for digital circuits. On one hand, this leads to very complex specifications that usually cannot cover all restrictions and potential uses of a certain block. On the other hand, different applications often require changes to the block's internal structure, so that a one-to-one reuse a is very rare event.

Thus, it is necessary to capture, store and retrieve design knowledge and not just fixed blocks. This knowledge has to be made available and applicable for new developments. It is represented by standardized design data, verbal information and design methods.

This means that extensive and rather informal documentation is a major part of the knowledge, which should at least be available by reference or even better it should be included in search keys. The *search criteria* have to be very flexible to support search for numerical ranges including handling of physical units, as well as mapping of semantically identical words onto each other. Neither straightforward full text search on the documentation nor strict database structures are suitable for this task.

Innovative approaches to solve this problem should also address possibilities to *re-enter the design* of a reused block at predefined points. To achieve this, the design flow must be analysed, and furthermore, standardized methods must be defined to store reuse information and to continue a design based on this information.

Besides the primary requirements stated above, several secondary requirements have to be deducted before an implementation can be started:

- In order to minimize the effort for the implementation of reuse techniques, it seems appropriate to use documentation and data that already appear in the existing design flow. This includes main results of the circuit development, a documentation of methods used for the development and references to at least one person that is responsible for the design.

- Formats for the representation of design results have to be defined. The most direct way is to use the formats of the tools that exist in the design flow. But this is not always possible, due to the huge amount of data involved and due to platform and license restrictions. Therefore, it is useful to derive additional preview documents in easy-to-display formats like ASCII, PostScript and PDF.

- A schema to store the captured information has to be chosen. For our approach, the standard file system is used, because this allows rapid prototyping and easy inclusion of all existing representations of design data. The main alternative would have been to define a database schema for a dedicated database management system.

- The basic way to search and access reuse data has to be selected. Here, the use of standard internet/intranet browsers with the corresponding traditional client/server architecture was preferred to more specialized implementations. This also minimizes the implementation effort. At the same time, it provides platform independence and remote access with zero overhead on the client side.

All these requirements lead to a set of tasks that would be ideally performed in a sequence. For the sake of flexibility and rapid prototyping, these tasks are concurrently handled by the DREAM prototype:
- Define the supported design flow and select the corresponding tools.
- Agree on a minimum set of structured documentation, prepared for possible extensions.
- Implement automated procedures to capture knowledge during the design flow.
- Schedule work to prepare the individual blocks for reuse.
- Implement a database with appropriate retrieval mechanisms.
- Collect design know-how of blocks that were designed before reuse was established.

9.3 IMPLEMENTATION

The DREAM prototype was implemented for use in the intranet of TEMIC Semiconductors and recently extended to whole ATMEL group to which TEMIC Semiconductors now belongs. Initially, input was collected from the design teams in separate IP surveys. From the resulting tables, one HTML file per block was automatically derived. Figure 9.1 gives an overview of the system structure.

Common browsers, like Netscape Navigator, are used for access and online entry of reuse blocks. This makes the data available on both UNIX and PC platforms.

The stored data comprises:
- Specification entries needed as keys to search in the reuse data.
- Additional information such as product codes and availability dates.
- Standardized references to separate intranet databases, e.g. to the contact person's data, and the fabrication process documentation.
- References to design data in Mentor Graphics format that can be accessed by click at sites where the appropriate license and file system is available.
- References to documentation in HTML, ASCII, PostScript, PDF are included where available. Typically, this is a preview of the design data or arbitrary information like data sheets, memos etc.

The search engine has been implemented using classical CGI techniques on the server side. The search form (see Figure 9.2) allows the user to enter free style key-

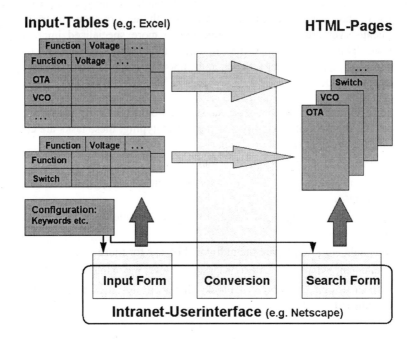

Figure 9.1 Structure of the DREAM implementation

words, numbers or ranges for predefined search criteria and for full text search. Option menus offer a selection of typical values for each search criterion. Client-side scripts offer dependent search criteria as soon as a circuit type ("function") is selected. Besides the implicit AND of different keywords, simple NOT and OR expressions are possible for each criterion.

The search result is a list of links to the HTML files for the matching blocks, which contain detailed information, including the afore mentioned links to referenced data. Figure 9.3 shows an example where a link to a PostScript preview has been activated.

Figure 9.4 shows a design re-entry using a link to a Mentor Design Architect database of the circuit. The corresponding design data are displayed read-only, but the user can start a new version branch using the normal framework functions. Server-side scripts handle version problems, such as attempts to access versions that are only available in external archives and new versions that have been designed after the reuse data were collected. Currently, the prototype supports Mentor Design Architect schematics and hierarchical ELDO/SPICE simulation data.

Besides the possibility to provide new entries in tables, an online entry form can be used by individual designers to enter new blocks or edit existing entries. Figure 9.5 gives an example.

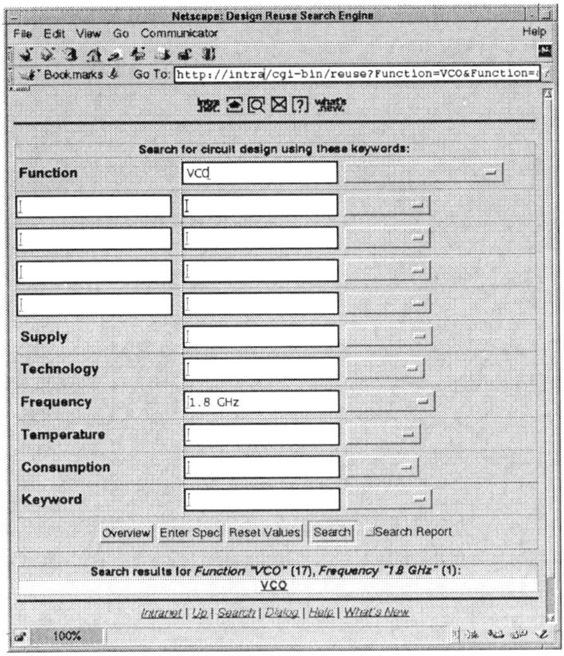

DESIGN REUSE EXPERIMENT FOR ANALOG MODULES "DREAM"

Figure 9.2 DREAM search engine

9.4 EXPERIENCE

The installation of the DREAM prototype is well accepted in the designer community as a possibility to find out about reusable circuit blocks and design know-how. Also the design management support the entry of design data and encourage the usage of the search engine. Currently, the database holds descriptions of more than 450 blocks, the majority purely is analog, the rest is mixed-signal and some digital designs are available. Because of the centralized data collection, the percentage of online inputs is still rather low. Also the number of blocks with references to design data suitable for re-entry is very low. Work is done to make the storage of blocks for reuse possible by menu entries within the usual design environment. Once these activities are completed, the number of online inputs is expected to rise.

In order to monitor the reuse activities, anonymous hit counts are performed for the search engine. By their nature, hit counts are very difficult to interpret, but it is sure that most potential users really tested the search engine in a short period after notice of availability or extensions. After that period the hit rates decreased to about 10 percent of those accesses per week, which is considered to be a solid user basis, taking into account the duration of typical design projects.

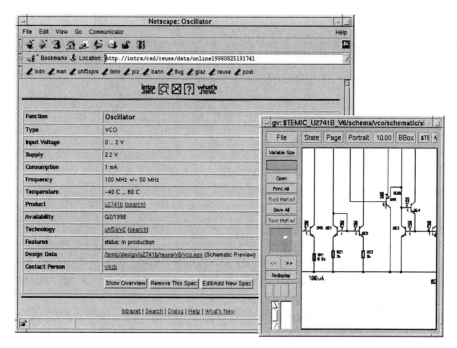

Figure 9.3 Data with circuit schematic preview

Moreover, the regular IC design reviews are being augmented by questions about blocks reused and blocks made available for reuse. Feedback from this action cannot be analysed yet.

9.5 FUTURE WORK

Currently, the integration of checkpoints for reuse data capturing into the design flow is a main activity. Other design data formats, especially for layout data will be defined soon. It has to be decided, how designers can be motivated to add new blocks without explicit survey activities. Maintenance of the database to avoid out-of-date information will also need increasing effort in the near future.

Besides this, work on the search engine is continued, e.g. to improve the mapping of identical keywords and acronyms and to allow more flexible matching of results while keeping the user interface as simple as possible. Also an "overview table" representation of search result will help to identify differences when a large number of blocks is found for a given search expression.

DESIGN REUSE EXPERIMENT FOR ANALOG MODULES "DREAM" 109

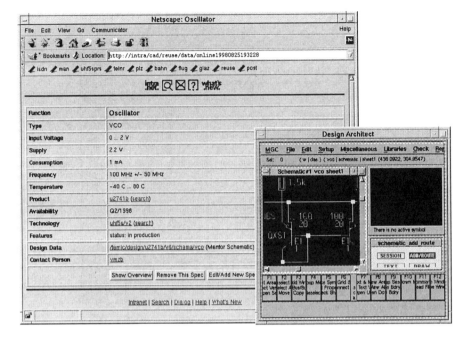

Figure 9.4 Data with design reentry window

9.6 ACKNOWLEDGEMENT

This work was supported by the German Bundesministerium für Bildung, Wissenschaft, Forschung und Technologie under contract 01 M 3037 C within the European MEDEA A409 project SADE (Systematic Analog Design Environment). The authors are responsible for the contents of this publication.

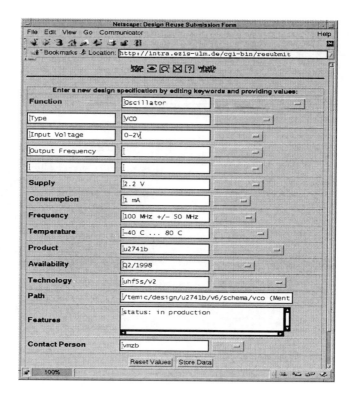

Figure 9.5 Online entry/editing of a design specification

Volker Meyer zu Bexten received his Diploma (1988) and Dr. rer. nat. (1994) degrees in Computer Science from the University of Dortmund, Germany. He wrote his diploma thesis at Siemens AG, Munich, Germany. From 1989 to 1995 he worked as a CAD research staff member at the Fraunhofer Institute of Microelectronic Circuits and Systems in Duisburg, Germany, where he also wrote his doctoral thesis. Since 1995 he is with TEMIC Semiconductors in Ulm, Germany, now as head of the CAD group. His research interests include layout synthesis for analog ICs and other areas that are related to computational geometry.

Anton Stürmer received his Diploma degree in Physics (1969) from the University of Würzburg, Germany. From 1969 to 1981 he worked as a research staff member at the Chemical Department of AEG-Telefunken research institute in Ulm in the development of technological processes. In 1981 he changed to the Design Center of Integrated Circuits with work in modelling of semiconductor devices and IC measurement. He is with TEMIC Semiconductors in Ulm, Germany, now as a member of the CAD group. His research interests include parasitic effects in analog ICs and the reuse of design knowledge.

10 REDESIGN OF AN MPEG-2-HDTV VIDEO DECODER CONSIDERING REUSE ASPECTS

H.-J. Brand*, R. Siegmund**, St. Riedel**,
K. Hesse**, and D. Müller**

*AMD Saxony Manufacturing GmbH
Dresden Design Center
Dresden, Germany

**Chemnitz University of Technology
Professorship Circuit and System Design
Chemnitz, Germany

10.1 INTRODUCTION

The well known gap between the increasing complexity of current IC designs and the productivity of available EDA tools requires the development of new design methods and approaches. This leads to the increased application of block-oriented development techniques using proven designs, cores or components as basic building blocks which is referred to as design reuse. Of course, design reuse techniques such as providing cores or module generators have already been applied for several years. But there is still a lack in supporting all aspects of design reuse during the complete design flow by appropriate tools. In order to comprehensively exploit the benefits of design reuse, two aspects have to be considered. Firstly, the design of reusable components and secondly, the design of new systems by reusing existing components.

Within a three-sided cooperation between the Heinrich-Hertz-Institut Berlin, the Fujitsu GmbH Frankfurt and the professorship circuit and system design at the TU Chemnitz an already existing implementation of an MPEG-2 video decoder [Krah96] (designed by the Heinrich-Hertz-Institut) was redesigned as single-chip solution using Fujitsu's sea of gate technologies. One of the major tasks of our department was the study of design reuse aspects in order to enable a future development of IPs using the MPEG-2 design as example. The main goals of this work are the evaluation of already existing approaches and tools as well as the formulation of basic guidelines which should be considered during the design process to support the design reuse.

The paper summarizes the experience we made during the design process with respect to the problems mentioned above. First we present some design reuse basics to give the background of our work. Then we explain the main features of both designs, the old and the new one and their consequences to the design process. In the following, we use two components of the video decoder to describe the experience we made for design by reuse and design for reuse, respectively. Finally, we summaries the conclusions from this work.

10.2 DESIGN REUSE

10.2.1 Intellectual Property

When developing complex systems, designers have increasingly to deal with the trade-off between a complete new design and the reuse of existing components. In many cases it will be cheaper to take existing components and cores and to assemble them with new ones to create a new system. The process of reuse and adaptation of existing components is called *design by reuse*.

One can distinguish between two sources of blocks for design by reuse. At first, there are components and cores from component libraries. They can consist of both components for fixed function and technology and of parametrizable, generic or generated components. Such libraries can be delivered by EDA or technology vendors and independent component providers.

The second way is the reuse of components from existing designs. System integrators have to decide for including available components based on their original specification and documentation. In case of changed specifications and constraints the components have to be adapted to it.

Independent of the source, it is implicated that the components have to be prepared for reusing them. This means that they must have comments in the source code, a meaningful documentation and well-defined standardized interfaces. Both, the component design and its interface to the design environment should include generics to guarantee a high flexibility and a wide range of possible applications.

As to be seen above there are some conditions for reusing components. In order to satisfy these terms many designers deal with the creation of reusable blocks. The creation of such proven components is named *design for reuse*. These components with a special functionality, standardized interfaces to the design environment and a comprehensive documentation have to be validated by verification. They need to be described using a proper VHDL subset applying a suitable modelling style. The VHDL models have to be tool-independent.

Furthermore, standardized components shall have a standardized pinout for compatibility. Standardized functions have to be realised by standardized design blocks which should be usable with the block specification. On-chip interfaces should meet the conditions set by the applicable standards (e.g. Virtual Socket Interface [VSIA97]). So the components are suitable for free combination with other design blocks from multiple sources.

Hard Cores (Hard VCs[a])	The vendor provides silicon layout (mask-level data) and the RTL simulation model. The component is fully implemented and verified. Function and properties are predictable, the core is difficult to modify and not portable. In case of generic IP components the tuning to process is low, and high if components are process specific.
Firm Cores (Firm VCs)	They consist of a RTL description of the component and physical placement data. The design has been floorplanned to achieve the basic design topology and synthesized into one or more technologies in order to obtain performance, area and power estimates. This representation is already an implemented form of the component, and therefore, portable only under limitations.
Soft Cores (Soft VCs)	Soft cores are shipped as a synthesized RTL description (VHDL, Verilog). The blocks are verifyable and have to be synthesized for implementation. It is the most flexible and portable form of a component representation, but its characteristics are hardly predictable and it cannot be easily protected against illegal use.

a. Virtual Components [VSIA97]

Table 10.1 IP component classification

Depending on the source such components are called *Virtual Components (VC)* or *Intellectual Property components (IP)*. An IP can be defined as an object or intangible item, whose major values come from the skill or artistry of its producer, not from the tangible medium of its delivery. IP components are, by definition, intellectual property [Lewi97]. Components sold as IPs have to be conform to design for reuse standards and guidelines, relating models, interfaces and documentation.

Besides the technical aspects there are legal ones if cores are sold to third-parties. The intellectual property rights of the designed components have to be protected. Marketing has to assure that the essential information related to the component are given to the integrators of IPs. Furthermore, it has to deal with quality assurance and management, and with guarantee aspects.

Depending on the implementation grade and the grade of the tool and technology dependency three types of components can be classified as shown in Table 10.1

Before shipping the components have to be validated. This is done via functional *verification* of the component models using testbenches. Stimuli sets have to be delivered for in-system verification of the integrated component. Efficient verification strategies and strategy combinations are applied to achieve reasonable run times. The models needed for the verification have to be delivered with the circuit descriptions for reusable components. Especially for complex components like an MPEG-2-decoder powerful verification strategies (e.g. cycle-based simulation, formal verification) can reduce the design effort considerably.

A last but not less important issue when creating IP components is a comprehensive *documentation* of the designed block. It has to be delivered by the component creator and in addition to the specification it is required for decisions related to the reuse of the existing design. It consists of both the knowledge of the designer and the special knowledge related to the actual design. It covers design decisions, *modelled and not-modelled properties* of the component or parts of it. The most important properties to document are the functional operation of the component, its parameters and operating modes, performance and timing requirements. The documentation has to cover such properties as the design size (estimated or back-annotated), power requirements, the acceptable operational ranges of temperature, the validity of the models and their limits as well as the accuracy of simulation results. Furthermore, the delivered information has to explain *the design* state of the component. Claims and assumptions need to be verified. The version history, known bugs and application notes need to be documented. It is possible to provide a system description, block diagram, register description, timing diagram and an explanation of the clock distribution. The documentation should also include bus interfaces, I/O configurations and a test description summary.

Using such a comprehensive documentation design-error risks are reduced significantly. The integrator of the components is able to apply valid parameters and possible optimizations of the component.

10.2.2 Reuse Techniques

As described in the previous chapter, components facilitated for design reuse may be fixed in functionality and technology which is surely appropriate for designs that are targeted at particular applications. However, components that are intended to be reusable for a wide range of applications need to be easily adaptable to new specifications that might include a change in functionality. In order to keep the modification effort small, reuse techniques should be employed such as *design parametrisation*, *component generation* or *object-oriented modelling*.

Design Parametrisation is a reuse technique that introduces parameters into an HDL design description in order to have certain degrees of freedom for the design functionality. The design is adapted to new functional specifications by assigning constant values to the parameters so that the design will be functionally "*specialised*". The advantage of this method is that no modifications will have to be made in the design description, as long as there is a set of parameter values that will adapt the design's functionality to the desired one. Yet, parametrisation of a design description also has some disadvantages: Implementation and verification effort for a parametrised design is considerably higher than for designs with fixed functionality. Moreover, all commonly used HDL's have their limitations in design parametrisability which are further tightened if the design description is to be syntheziable.

While design parametrisation is a reuse technique well suited for higher levels of design abstraction, for lower levels **Component Generation** may be used. A generator produces from a unique and consistent formal specification a low-level component description, e.g. a netlist, which is optimised in area, performance and power consumption for a certain technology. Component generation overcomes the limitations of HDL design parametrization and is mainly used for standard datapath components with a high amount of regularity such as arithmetic modules.

Object-oriented modelling is an emerging reuse technique for hardware design and is extensively used in the software domain to facilitate reuse of code blocks. Hardware components are considered as objects with a set of properties and a set of methods to change these properties. The two basic concepts of object-oriented modelling, *polymorphism* and *inheritance*, provide a means to adapt the design's functionality to a new specification: From an existing design object new objects may be derived. New properties and methods may be added to the design objects and inherited methods may be redefined. To give an example, one may think of a binary counter which is characterised by the property 'counter value' (which is in fact a counter variable) and a method to increment the counter value. If a grey-code counter is needed, a new object may be derived from the counter object and the method that increments the counter value may be redefined to gray-code increment. The example shows that object-oriented modelling is best suited for high-level algorithmic design descriptions, but considerable effort is put in extending this technique to lower levels of design abstraction. In fact, VHDL has concepts of object-oriented modelling implemented, e.g., the overloading of arithmetic operators for different argument types. Objective VHDL is an attempt to provide a formal language for object-oriented modelling of hardware design.

10.3 REDESIGN OF AN MPEG-2-HDTV VIDEO DECODER

10.3.1 MPEG-2-HDTV Video Decoder

The two chips BISTRO and MOFA developed at Heinrich-Hertz-Institut Berlin within the HDTV$_T$[1]-project were the basis of this design. These chips were designed using a ES2 0.7 m-CMOS-standard cell library with chip frequencies of 67,5 MHz (BISTRO) and 54 MHz (MOFA) [Krah96]. The new single chip video decoder called HiPEG itself is designed in a 0.35 m-sea of gate technology with embedded RAMs called CE61 and it is clocked with 108 MHz chip frequency.

HiPEG is a configurable video decoder for interlaced HDTV (1920x1152 pixel/ 50 Hz or 1920x1088 pixel/60 Hz) or progressive (non-interlaced) SVGA sequences jointly designed at the Heinrich-Hertz-Institut and at the TU Chemnitz. Existing video decoders are only able to decode standard TV (720x576 pixel) or H14 (1440x1152 pixel) [Krah96]. HiPEG can read a transport stream as well as a packetized elementary stream with a data rate of 80 Mbit/s. The output is programmable for a digital output stream with YUV according TV norm CCIR 601, HDTV norm according ITU-R BT or with RGB output. It is possible to configure the ports to external SDRAMs of frame buffer. Figure 10.1 shows the architecture of HiPEG. [Schl97]

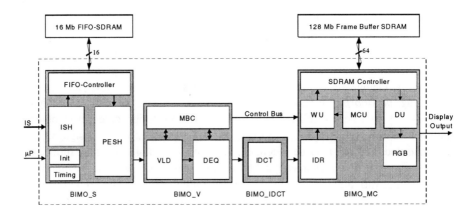

Figure 10.1 HiPEG architecture

---

1. Hierarchical Digital TV Transmission

10.3.2 Design Reuse Aspects

Basic goal of the single chip implementation was to reuse as many components from the existing design as possible. Reused should be the behavioural VHDL-specifications as well as the verification environment (testbenches, C reference model, accuracy checks etc.)

Besides the mapping to a new technology, some functional changes were required (such as HDTV capability) leading to a modified system specification. Consequently, it was necessary to analyse first if the reuse and adaptation effort will be considerably lower than a complete new design. In order to support this analysis task we distinguish between the following types of specification changes (examples of HiPEG are given in parentheses):

- *functionality* (remove SNR and Spatial Scalability, new RGB output)
- *interface* (interface between motion compensation and the rest of the chip is now within chip, external frame buffer is an SDRAM instead of SRAM)
- *technology* (synchronous internal RAMs instead of asynchronous ones)
- *constraints* (clock frequency)
- *design flow* (using the module generator CGGEN [Bran97] for the implementation of arithmetic components)

As a consequence of the variety of changes each HiPEG component was separately analysed. Besides the question for changes, it has to be examined which effort these changes require. From our experience, the following requirements turned out to be important for the effort related to the reuse of a design:

- *modelling* - available sources and models
- *verification* - available testbenches and verification environments etc.
- *documentation* - component and design flow documentation
- *design support* - scripts for automating the design flow, project data structure, version control etc.

Obviously, the analysis of the *reusability* (comprising the change request summary and the examination of the consequent design effort) of the various components of HiPEG leads to different results. Hence, for some components it was appropriate to reuse them whereas for others the creation of a new design requires less effort than the adaptation of existing designs.

In the following two sections two components are used as example to illustrate the varying approaches applied. Firstly, the design of the inverse quantiser will be described giving an example of a component, which was reused. In Section 3.4 the IDCT design is explained. This component is a complete new design, but should be a reusable component or IP.

10.4 DESIGN BY REUSE (INVERSE QUANTISER)

As mentioned in the previous section, the major goal was the reuse of the existing designs. The requirements of the new HiPEG chip cause the following changes in the inverse quantiser specification with respect to the types of specification changes defined in section 3.2.

- *functionality*: no changes
- *interface:* no changes
- *technology*: new technology provides synchronous instead of asynchronous RAMs
- *constraints*: new clock frequency of 108 MHz
- *design flow*: new design environment, module generator CGGEN available

The modification of the control circuitry for the internal RAM and the optimization of arithmetic calculations by removing several parallel multiple-clock-cycle multipliers by single high-performance multipliers produced with CGGEN were identified as the most expensive design tasks during the redesign.

With respect to the reusability requirements the following statements could be made.

- *modelling* - VHDL sources had to be changed (modelling style, guarantee tool independence, CGGEN requires special package for arithmetic operators)
- *verification* - existing testbenches could be mostly reused
- *documentation* - existing documentation mainly describes the implementation, no design flow documentation available
- *design support* - scripts for automating the design flow were not available

As a consequence, the inverse quantiser turned out to be a component with reasonable change effort caused by the new technology and design environment. Thus the original design was reused and changed according to the new specification and constraints. In [Sedn97] the process of analysis and redesign is described in detail.

In order to minimize the design effort the inverse quantiser was not redesigned in such a way that it meets the requirements for an IP. Owing to the optimizations mentioned above an area reduction of about 15% was achieved for the inverse quantiser.

10.5 DESIGN FOR REUSE (IDCT - INVERSE DISCRETE COSINE TRANSFORM)

Table 10.2 summarizes the main constraints for the old and the new IDCT design.

constraints	old design	new design
clock rate	54 MHz	108 MHz
technology	0.7 m	0.35 m
internal RAM	asynchronous	synchronous
design flow	tools from one provider used	tool independent

Table 10.2 IDCT features

Both designs use a two-pass computation for the two-dimensional IDCT using a one-dimensional IDCT as computation core. Hence, an internal memory is required for a row-to-line conversion between the first and second pass of the computation.

Due to the limited performance of the ES2 technology the arithmetic core of the original IDCT design had to be parallelized. In general four identical arithmetic blocks were used to calculate four results per clock cycle in order to achieve the necessary throughput. A similar problem had to be fixed for the internal RAMs used as row-to-line converter. The access time exceeds the clock period requiring two clocks for one memory access. As a consequence this memory was splitted into two separate blocks.

The new design could be implemented using a faster technology supporting a one clock cycle memory access even for the required clock rate of 108 MHz. But in contrast to the ES2 technology Fujitsu's technology provides synchronous internal RAMs requiring a modified RAM control procedure. With respect to the arithmetic core the module generator CGGEN for producing high performance arithmetic components could be applied. Due to the better area-performance characteristics, which can be achieved compared to commercially available synthesis tools CGGEN enables the implementation of a considerably smaller and faster IDCT arithmetic core than in the original design. So all parallel components could be removed. The new arithmetic core calculates two results per clock cycle. The area of the design could be reduced from about 60 k gates to about 25 k gates.

To completely exploit the performance of the technology used a new IDCT architecture had to be implemented making a reuse or redesign of the original IDCT ineffectively. In contrast to the design itself the IDCT testbench could be reused because the IDCT interface was only slightly changed.

Additionally two parameters were introduced into the VHDL description in form of generics. Table 10.3 summarises the IDCT parameter set.

The first parameter N allows the configuration of the IDCT to transform pixel blocks of any arbitrary size NxN under the condition that N must be even. This parametrisation was based on the observation that for even N the 1D-IDCT computation core structure does not change. A perl script is used to compute all N IDCT transformation coefficients which are stored as constants in a separate VHDL package, and to generate the parameter strings for the arithmetic module generator CGGEN. A problem that is common for parametrised HDL design descriptions was found to be the internal RAM cell. The storage size of this cell clearly depends on the parameter N. Currently, no commercial synthesis tool can automatically infer technology-dependent RAM cells from an HDL description so that this cell has to be manually instantiated in the design.

parameters	old design	new design
size of pixel block	fixed to 8x8	NxN (N even)
No. of pipeline stages in 1D-IDCT computation core	fixed to 2	M

Table 10.3 IDCT parameter set

The second parameter M was introduced to vary the number of pipeline stages in the 1D-IDCT computation core. The combinational logic depth of this core, which contains a large amount of arithmetic modules such as adders and multipliers, eventually determines the maximum clock rate of the IDCT. Moreover, an increase of parameter N will generally lead to a higher logic depth in the computation core and hence the maximum clock rate will be reduced. This effect is compensated through insertion of an appropriate number of pipeline stages into the computation core. To do this, the VHDL design description was modified such that during synthesis a number of M register banks will be attached to the output of the 1D-IDCT computation core and then the core netlist is retimed to equalise the combinational logic depth between the pipeline stages.

The new IDCT design should be a reusable component. In order to guarantee this requirement the following aspects were considered during IDCT design.

(i) modelling

The behavioural specification uses a specially defined VHDL subset to achieve tool independence (for simulation as well as synthesis). This includes the use of the VHDL-package modelling the behaviour of the arithmetic components generated with CGGEN.

Additionally, some special features were modelled in order to enable the parametrization of the IDCT to achieve a high flexibility. So the internal IDCT structure can be changed (may be by faster components or calculation algorithm modifications for the arithmetic core) without requiring changes in the interface timing. This could be achieved by producing the validity signal for IDCT output data from separate valid signals of the internal components. Moreover, an additional output was added to the IDCT denoting the address of a pixel within a block of pixels. Therefore, the internal order of calculation can be changed without influencing following components.

Apart from the standard VHDL model a VHDL+ model was developed [Blum97], [ICL97]. VHDL+ is a superset of VHDL and incorporates the concepts of the interface-based design methodology as proposed in [Rows97]. This methodology separates the functionality of system components from their communications (e.g. data protocols) in an orthogonal fashion, and hence, enables the separate specification and implementation of the system components and the communication hardware. Communication between system components is modelled by means of interfaces that contain protocol specifications, while system components may use high-level messaging to exchange data. The interface itself specifies all levels of protocol abstraction from high-level message tokens down to low-level signalling, and therefore enables communication between system components which are modelled at different levels of abstraction ([Sieg98b], [Sieg98a]). Figure 10.2[1] illustrates the concept of interface-based system modelling.

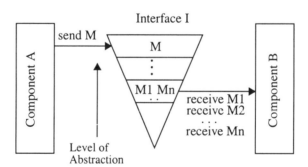

Figure 10.2 Interface-based design

Component A uses *interface I* to communicate with *component B* by sending messages M. *Component A* may contain an abstract specification of its functionality

---

1. from: SuperVISE: System Specification and Design, Methodologies. ICL Manchester, 1998.

and uses high-level messaging ("send M") while *component B* might be specified in more detail and is therefore capable of composing message M according to the protocol rules of *I*. The interface automatically performs translation of messages between levels of hierarchy according to the interface specification and thus enables design units which are at different levels of abstraction to communicate with each other.

According to interface-based modelling, the VHDL+ simulation model of the IDCT consists of two primary units: the abstract functional model which contains the IDCT computation algorithm and an interface description with the protocol specification for data transfers to and from the IDCT. The VHDL+ model of the IDCT is well suited as simulation model in terms of IP: It hides implementation details while being functionally correct and cycle accurate, and it enables the IP consumer to adapt the interface specification to its own systems requirements without modification of the functional model. Currently, a change of the VHDL+ interface specification of the IDCT model requires the manual change of the interface of the synthesis model. To overcome this problem, research is under way to synthesize the interface of an IP component according to the protocol rules in the interface specification.

(ii) verification

The existing testbench could be reused with minor changes. This testbench includes the accuracy check according the CCITT standard.

(iii) documentation

The documentation includes source code comments as well as a design and design flow documentation (design environment, modelled and not-modelled properties, solutions for special problems with respect to behaviour, synthesis etc.)

(iv) design support

The new IDCT design was carried out within a global project data structure using RCS version control. This data structure is used for all components developed at the TU Chemnitz and supports the easy management of release creation necessary for data exchange between the two partners involved in the design. Script files for automating the design flow were developed simultaneously or can be automatically produced (e.g. for compiling VHDL sources) from a so-called hierarchy file describing the structure of the according component.

10.6 SUMMARY

In summary we came to the following conclusions from our experience with the redesign of the MPEG-2-HDTV video decoder.

We were able to reuse less components as expected (in principle only the variable length decoder and the inverse quantiser). The main reasons for that were the changes in functionality caused by the HDTV features, the inadequate documentation of the existing designs and their close adaptation to the originally used design tools and technology. In contrast to the designs most of the test and verification environment could be reused requiring only a limited change effort.

The decision whether to reuse an existing design or not requires a detailed effort analysis. From our point of view the following aspects should be considered for evaluating the reusability of a component:

- Extent of specification changes (functionality, interface, technology, constraints, design flow)
- Does the component meet the requirements for reusability (modelling aspects, verification support, documentation, design support)?

The verification was one major problem with which we were confronted. When considering the needs of the specific problems of the MPEG-2 video decoder, there are different requirements for component and system verification, respectively. Components have to be tested using testbenches. This allows the verification of the correctness of the results produced by the components when stimulating via test vectors. The system verification, however, is performed by estimating the quality of test pictures and sequences (e.g., using conformance-bitstreams defined in the MPEG-2 standard). There are permissible limits for estimating the system function as correct. A full covered verification for such a system is too complex and therefore impossible.

First prototypes of HiPEG have been delivered in April 1998 and were tested thoroughly. The chip worked satisfactory, it could decode special test data streams as well as real-world DVB (Digital Video Broadcast) data streams in all required resolutions up to HDTV. Minor flaws occurred in some of the rather exotic display modes and when the chip was faced with non-conforming data streams (which we learned happen to appear with some broadcasters). Currently, an update of the design is under development, which will not only fix these bugs but also implement additional features. It is intended to have this new chip ready in Q1 of 1999.

Hans-Jürgen Brand received both his M.S. (1989) and Ph.D. (1994) degrees in Information Technology from the Chemnitz University of Technology. From 1994 to 1997 he worked as Assistant Professor at the same university. He currently is a Senior Digital Architect at the AMD design center in Dresden. His research interests include system verification, system level design for communication products and module generation.

11 REUSE CONCEPTS IN GROPIUS

D. Eisenbiegler and C. Blumenröhr

Institute for Circuit Design and Fault Tolerance (Prof. Dr.-Ing. D. Schmid)
University of Karlsruhe, Germany

11.1 ABSTRACT

This paper introduces our new hardware description language named Gropius[1]. Gropius covers all abstraction levels from the gate level to the system level. Design reuse and abstraction are supported in a systematic manner. Gropius was designed for a formal synthesis scenario, where synthesis is performed by applying mathematical derivation steps, thus guaranteeing correctness of the synthesis process. Since Gropius was defined in a mathematical manner, its semantics is precise and unambiguous. In the paper special attention is focused on the reuse concepts that are realized in Gropius.

---

1. This work has been partly financed by the Deutsche Forschungsgemeinschaft, Project 623/6-1

11.2 INTRODUCTION

In circuit design abstract design levels are more and more used in order to manage the complexity of today's systems. Synthesis means mapping an abstract circuit description given by the designer to a concrete implementation in the real world. In synthesis, hardware description languages play an important role. They determine the set of circuit descriptions to be considered as input and output of some synthesis step. A synthesis concept is always closely related to some specific hardware description language. Therefore, the design of a good hardware description language is very important.

What are the main objectives when designing an efficient hardware description language? We believe, that there are three major points that have to be considered:

- **expressiveness** To get along with the complexity of large circuits, one has to go beyond pure gate level net lists. More abstract levels of circuit descriptions are required: rt-level, algorithmic level, system level. At these levels among other features, hardware description languages have to support abstract data types as well as abstract timing descriptions. The features provided by the hardware description language must correspond to a design concept that allows the circuit designer to represent circuits in a compact manner. Providing a good concept for design reuse is a major contribution to the quality of a hardware description language.

- **unambiguous semantics** It seems to be clear, that it is necessary to explain what the syntactical elements of a hardware description language stand for. Good hardware description languages are always related to a simple semantics that can easily be described in a mathematical notation. A hardware description language with a complicated semantics based on an awkward timing model leads to design errors due to misunderstandings.

- **minimum size** A hardware description language should be as small as possible. This means strictly eliminating redundant constructs. Both the size of a hardware description language and the complexity of the semantics determine the costs for teaching it to new circuit designers.

We believe that existing hardware description languages such as VHDL or Verilog are far worse than they could be. Most problems of hardware description languages can be illustrated with VHDL. VHDL is one of the most widespread hardware description languages. In some sense VHDL is very expressive: the set of systems one can describe with VHDL goes beyond what may be built in reality. Therefore, all synthesis tools have to reduce themselves to subsets of VHDL. There is an informal semantics for VHDL [IEE96], however it is not unambiguous. Several attempts have been made to give VHDL a formal, precise semantics, however, there is no common standard [Kloo95]. As a result, different tools interpret VHDL sources in different ways. In general, tools with VHDL interfaces are not compatible, designs cannot be reused and it is not possible to process VHDL code from one tool to the next. Besides the semantics problem of VHDL, there is also a lack of the expressiveness of the language. The design concepts of VHDL are pretty poor compared to modern programming languages in software design. Several attempts have been made to extend VHDL by modern design/reuse concepts [Rade97, Swam95].

11.3 GROPIUS - A SURVEY

This article presents a new hardware description language named Gropius[1]. Gropius has an exact, unambiguous mathematical semantics. It was designed for a formal synthesis tool, where synthesis is performed within a theorem prover system by applying mathematical rules (see [Kuma96] for a survey on formal synthesis). All constructs of Gropius have been defined within logic - in other terms: Gropius is nothing but a subset of higher order logic. Gropius is pretty expressive and supports different techniques related to design reuse. First we will briefly introduce Gropius and then explain the techniques for design reuse.

Gropius is a functional hardware description language. It provides means for the user to define arbitrary functions. User defined functions are the common basis for hardware descriptions at different levels of abstraction (see section 11.4). Gropius is strongly-typed and, unlike Ella [Mori93], includes an automatic type inference mechanism. Gropius supports polymorphic and generic structures and allows parameterizing circuits with subcomponents. These are very powerful means for a systematic reuse of designs. Unlike HML [Lear93], Ruby [Shar95], DDD [John91] or Lustre [Halb91], Gropius is not restricted to lower levels of abstractions, but also supports circuit descriptions at the algorithmic and system level.

Gropius is a very small language with a minimum number of constructs. There are only 11 syntax rules (see figure 11.1). Also the number of basic key words is pretty small: there are 5 basic boolean operators, 25 abstract operators for non-boolean expressions (polymorphism, enumeration types, arrays etc.), 9 interface patterns and 9 different k-processes (see table11.1). The small size makes it easy for the user to learn the language. In VHDL for example, there are more than 150 syntax rules (see [IEE96]).

Gropius is designed to support a specific design methodology. Design reuse is one of its strengths. In this paper, we will give a brief introduction to Gropius and describe some major concepts and their significance as to design reuse.

11.4 DESIGN REUSE ACROSS ABSTRACTION LEVELS

In Gropius, all functions are based on a fixed set of elementary boolean and abstract operators (see figure 11.2). User defined functions are subdivided into three groups: boolean dfg-terms, dfg-terms and p-terms. Boolean dfg-terms are a subset of dfg-

---

1. Walter Gropius (1883-1969), founder of the Bauhaus (form follows function)

$$
\begin{array}{rcl}
vblock & ::= & variable \mid "(" \; \{ \; vblock \; "," \; \} \; vblock \; ")" \\[4pt]
expr & ::= & variable \mid constant \mid "(" \; \{ \; expr \; "," \; \} \; expr \; ")" \mid \\
 & & operator \; "(" \; expr \; ")" \\[4pt]
\textit{dfg-term} & ::= & "\lambda" \; vblock \; "." \quad \{"\text{let}" \; vblock \; "=" \; expr \; "\text{in}"\} \; expr \\[4pt]
\textit{sequential circuit} & ::= & "\text{automaton}" \; "(" \; \textit{dfg-term} \; "," \; expr \; ")" \\[4pt]
block & ::= & "\text{PARTIALIZE}" \; \textit{basic\_block} \mid "\text{WHILE}" \; condition \; block \mid \\
 & & block \; "\text{THEN}" \; block \mid "\text{IFTE}" \; condition \; block \; block \mid \\
 & & "\text{LOCVAR}" \; constant \; block \mid \\
 & & "\text{LEFTVAR}" \; block \mid "\text{RIGHTVAR}" \; block \\[4pt]
program & ::= & "\text{PROGRAM}" \; constant \; block \\[4pt]
\multicolumn{3}{l}{\textit{algorithmic dfg-circuit description} \; ::=} \\
 & & \textit{dfg-interface pattern} \; "(" \; \textit{dfg-term} \; "," \; \textit{number of cycles} \; ")" \\[4pt]
\multicolumn{3}{l}{\textit{algorithmic p-circuit description} \; ::=} \\
 & & \textit{p-interface pattern} \; "(" \; program \; ")" \\[4pt]
\textit{s-interface} & ::= & "(" \; "\text{reset}" \; \{ \; "," \; channel \; \} \; ")" \\[4pt]
\textit{s-process} & ::= & \textit{algorithmic dfg-circuit description} \mid \\
 & & \textit{algorithmic p-circuit description} \mid \\
 & & \textit{k-process} \\[4pt]
\textit{s-structure} & ::= & "\exists" \; channel \; \{ \; "," \; channel \; \} \; "." \\
 & & \textit{s-process s-interface} \; \{ \; "\wedge" \; \textit{s-process s-interface} \; \}
\end{array}
$$

Figure 11.1 Syntax of Gropius

terms and dfg-terms are a subset of p-terms. The set of p-terms covers the entire set of computable functions. Dfg-Terms are non-recursive compositions of elementary operators. Boolean dfg-terms are exclusively build on basic boolean operators.

Unlike most standard hardware description languages such as VHDL and Verilog, Gropius is strictly divided into sub languages each corresponding to a specific abstraction level: Gropius-0 - gate level, Gropius-1 - rt-level, Gropius-2 - algorithmic level, Gropius-3 - system level. Due to the common core of the sub languages, Gropius is not just a set of hardware description languages. The elementary constructs and the user defined functions are a common starting point for all circuit descriptions.

Boolean dfg-terms and dfg-terms are an appropriate means for representing combinatorial circuit descriptions at the gate level and at the rt-level, respectively. The operator **automaton** is used for representing sequential circuits. Given some dfg term f representing the combinatorial part of the sequential circuit and some ini-

boolean operators		AND, OR, INV, T, F
abstract operators	- *polymorphism*	MUX, EQ
	- *enumeration types*	enum, next
	- *arrays*	mkarray, spread, pick, modify, cut, append, shift, rev, comb, split, shrink, unshrink, ripple
	- *one element type*	one
	- *optional values*	none, any, CASE_option
	- *variant records*	INL, INR, CASE_sum
dfg-interface patterns		dfg_interface_cycle, dfg_interface_start, dfg_interface_reset, dfg_interface_pipeline, dfg_interface_system
p-interface patterns		p_interface_cycle, p_interface_start, p_interface_reset, p_interface_system
k-processes		Synchronize, Double, Join, Split, Fork, Combine, Choose, Sink, Counter *n*

Table 11.1 Basic constructs of Gropius

tial state q, automaton maps (f,q) to a sequential circuit automaton(f,q). automaton is used both for gate level circuits and rt-level circuits (see [Eise97a]).

Dfg-terms are not only used as means for representing combinatorial parts of circuits, but are also used for describing i/o relations at the algorithmic level. At the algorithmic level the circuit designer may use arbitrary computable functions as starting point. In Gropius, however, p-terms and dfg-terms are strictly divided. For specific synthesis tasks, it is worthwhile knowing, that the function to be considered is a pure data flow graph and - other than with p-terms - non-termination is not a matter that has to be considered.

Besides an i/o relation also an interface description must be given for performing synthesis at the algorithmic level (Gropius-2). A fixed set of interface patterns is used to describe how the circuit implements some function g. They differ in the way the circuit communicates with the environment. There are two classes of interface patterns, which are either related to dfg-terms or to p-terms. The combination

130   REUSE TECHNIQUES FOR VLSI DESIGN

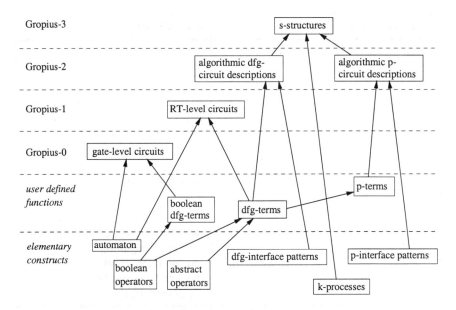

Figure 11.2 Gropius

of some p(dfg)-interface pattern with some p(dfg)-term results in an algorithmic circuit description that is the starting point for high-level synthesis.

At the algorithmic level, only single processes are considered. Gropius also supports multi-process systems at the system level (Gropius-3). In Gropius-3, systems are represented by structures of processes. In Gropius-3 there are two kinds of processes: algorithmic circuit descriptions and k-processes. Processes in Gropius-3 use a fixed communication scheme: higher order petri nets [Jens92], i.e. petri nets with marks having values. Therefore only two of the above mentioned interface patterns are allowed in Gropius-3: dfg_interface_system and p_interface_system. The k-processes are used for communication purposes: delaying marks, duplicating marks, synchronizing marks etc.

In Gropius, synthesis means translating circuit descriptions from Gropius-3, Gropius-2 and Gropius-1 down to Gropius-0. In Gropius, all circuit descriptions are synthesizable, i.e. can be mapped to an equivalent Gropius-0 representation. Annotation: for generic circuit description a concrete instantiation has to be made and for abstract data an encoding is required. An example concerning synthesis at the rt-level can be found in [Eise97b], a description, where high-level synthesis is performed, is given in [Blum98], and for system level synthesis see [Blum99].

In the following sections, we go into the details of those reuse concepts that are supported in Gropius.

11.5 EVERYTHING CAN BE ABBREVIATED

Gropius is pretty easy to use. The user may build an arbitrary new expression and give the expression a new name. For example, the user can define new functions:

nor(a,b) := not (or (a,b))

Such a definition means an extension to Gropius. From now on nor may be used as an additional function. The semantics is obvious: nor(a,b) is nothing but an abbreviation for not (or (a,b)).

Gropius is a higher order programming language. Besides functions, the user can, for example, also define new control structures, which are functions of functions. Gropius provides only a minimum number of control structures. Although there is only a WHILE-loop and the REPEAT loop is not provided, the REPEAT loop can easily be defined by the user:

NOP := PARTIALIZE ($\lambda x. x$)

LOOP A c B := A THEN (WHILE c (B THEN A))

REPEAT A c := LOOP A c NOP

It has to be noted, that all free variables on the right hand side of a definition must be parameters of the newly defined construct. All operators on the right hand side must already have been defined. The equation must be non-recursive, i.e. the function being defined must not appear on the right hand side.

Since the definition of new functions and control structures is an extension to Gropius, the designer can easily reuse such abbreviations in different designs.

11.6 POLYMORPHISM

Gropius supports polymorphism. There are two basic polymorphic functions: a multiplexer MUX with one control bit and a variable signal type, and an equivalence gate EQ whose two input values are of the same but variable type. Based on these two elementary polymorphic functions, the user may derive more complex polymorphic functions such as:

MUX4(r,s,a,b,c,d) := MUX(r,MUX(s,a,b),MUX(s,c,d))

Polymorphic functions can be used in different instantiations. The function MUX, for example, has type bool $\times \alpha \times \alpha \rightarrow \alpha$, where α is a type variable. α can be instantiated in an arbitrary manner. This need not be done explicitly, but is derived by the automatic type inference mechanism from the context in which MUX is used. In the expression MUX($a,$T$,b$), for example, α becomes bool, whereas in MUX($a,$(T$,b$),($c,$OR(d,e)) the type variable α is instantiated with bool $\times$ bool.

The advantage is that the same function can be reused in different designs realizing different components by differently instantiating the types. In contrast to Gropius, the type system of the language Ella, which also is a functional language and allows user defined types, is not polymorphic.

11.7 PARAMETERIZATION WITH CIRCUITS

Since Gropius is a higher order programming language, which means that functions can be passed as arguments, returned as values and stored as variables, it is allowed in Gropius to have circuits as parameters of circuits. The following, user defined function quad maps some gate f to a structure consisting of four f gates.

quad f (a,b) := ($f(f(a))$, $f(f(b))$)

In this example f is polymorphic. Its type is $\alpha \to \alpha$. The construct quad is a very general pattern representing a simple structure of four equal combinatorial circuits in a very general manner: f may be any function of type $\alpha \to \alpha$ with α being an arbitrary type. It is evident that this is a very powerful means for the reuse of designs. Once a circuit is defined, it can be reused in another design as a component of a more complex circuit. Together with the fact that circuits can be described polymorphic, very different circuits can be derived from such general patterns.

11.8 REGULARITY

Gropius supports regularity in a systematic manner. There is one basic regular structure named ripple. ripple is a generic, polymorphic structure of a sequence of circuits (see figures 11.3 and 11.4). Regular circuit structures lead to regular signal bundling. In Gropius, arrays are used for representing regular signal patterns. Besides ripple, Gropius provides a set of functions on arrays (see table 11.1). Figures 11.3 and 11.4 give an impression on how powerful complex structures can be described in Gropius.

The definitions are build bottom up from the Gropius basics. Some of the corresponding structures are sketched in figure 11.4. As can be seen, the user can start defining some general, polymorphic patterns such as ripplec and ser and later on instantiate them to derive more concrete circuits such as ANDN. One just has to instantiate the parameter n to achieve a concrete, realizable circuit. (ANDN 7) for example is an AND-gate with 7 inputs.

11.9 STRICT SEPARATION BETWEEN FUNCTIONAL AND TEMPORAL ASPECTS

At the algorithmic level (Gropius-2) as well as on the system level (Gropius-3), the circuit descriptions consist of two components: a functional i/o description and an interface pattern. The functional part only describes, how some inputs are mapped

$$
\begin{aligned}
\text{EQP } m\, n &:= \text{EQ}(\text{enum } (m+n+1)\, m, \text{enum } (m+n+1)\, n) \\
\text{first } n\, x &:= \text{pick } n(\text{enum } n\, 0, x) \\
\text{maxenum } n &:= \text{enum } n\, (n-1) \\
\text{last } n\, x &:= \text{pick } n(\text{maxenum } n, x) \\
\text{last' } n\, (x,y) &:= \text{MUX}(\text{EQP } n\, 0, y, \text{last } n\, x) \\
\text{constantly } f\, i &:= f \\
\\
\text{ser } n\, f\, x &:= \text{last' } n\, (\\
&\quad \text{SND}(\text{ripple } n\, (\lambda i.\, \lambda(a,b).\, (\text{one}, f\, i\, a))\, (x, \text{spread } n\, \text{one})),\\
&\quad x\\
&\quad)\\
\text{par } n\, f\, x &:= \text{FST}(\text{ripple } n\, (\lambda i.\, \lambda(a,b).\, (f\, i\, b, \text{one}))\, (\text{one}, x))\\
\text{for } n\, f &:= \text{par } n\, (\lambda i.\, \lambda x.\, f\, i)\, (\text{spread } n\, \text{one})\\
\text{rippleb } n\, f\, (a,b) &:= \text{let } (c,d) = \text{ripple } n\, f\, (a,b) \text{ in } (c, \text{last' } n\, (d,a))\\
\text{ripplec } n\, f\, (a,b) &:= \text{last' } n\, (\text{SND}(\text{ripple } n\, (\lambda i.\, \lambda x.\, (\text{one}, f\, i\, x)))\, (a,b), a)\\
\text{rippled } n\, f\, (x,d) &:= \text{MUX}(\text{EQP } n\, 0,\ d,\\
&\quad \text{MUX}(\text{EQP } n\, 1,\ \text{first } n\, x,\\
&\quad \text{ripplec } (n-1)\, f(\text{first } n\, x, \text{shift } 1\, (n-1)\, x)\,)\,)\\
\text{rect } m\, n\, f\, (a,b) &:= \text{ser } m\, (\lambda i.\, \lambda(x,y).\, \text{swap}(\text{rippleb } n\, (f\, i)\, (x,y))\, (a,b))\\
\text{MUXN } n\, (c,a,b) &:= \text{par } n\, (\text{constantly}(\lambda(s,(a,b)).\, \text{MUX}(s,a,b)))\\
&\quad (\text{comb } n\, (\text{spread } n\, c, \text{comb } n\, (a,b)))\\
\text{ANDN } n\, x &:= \text{rippled } n\, (\text{constantly AND})\, (\text{T}, x)\\
\text{ADDERN } n\, (cin, a, b) &:= \text{rippleb } n\, (\text{constantly}(\lambda(c,(x,y)).\text{FADD}(c,x,y)))\\
&\quad (cin, \text{comb } n\, (a,b))\\
\text{EQN } n\, (a,b) &:= \text{ANDN } n\, (\text{par } n\, (\text{constantly EQ})\, (\text{comb } n\, (a,b)))\\
\text{MUXT } n\, (s,a) &:= \text{first } 1\, (\\
&\quad \text{SND}(\text{ser } n\, (\lambda i.\, \lambda(s', a').\\
&\quad\quad \text{let } (x,y) = \text{split } 2^{n-(i+1)}\, (\text{shrink } 2^{n-(i+1)}\, a')\text{ in}\\
&\quad\quad (s', \text{MUX}(\text{pick } n\, (\text{enum } n\, i, s'), y, x))\\
&\quad)(s,a))\\
&\quad)
\end{aligned}
$$

Figure 11.3 Example I of derived structures

to some outputs. To bridge the gap between the higher abstraction levels and the rt-level, an interface behaviour has to be defined additionally, which describes, how the circuit communicates with its environment.

In most approaches, the functional and interface description can be mingled in an arbitrary manner. In our approach however, they are strictly separated. Whereas at the system level a fixed interface behaviour is defined, at the algorithmic level one can choose from a fixed set of interface patterns. This means that at the algorithmic level the designer can first write some ordinary, time independent program and can then select one of the interface patterns to define the way the circuit communicates

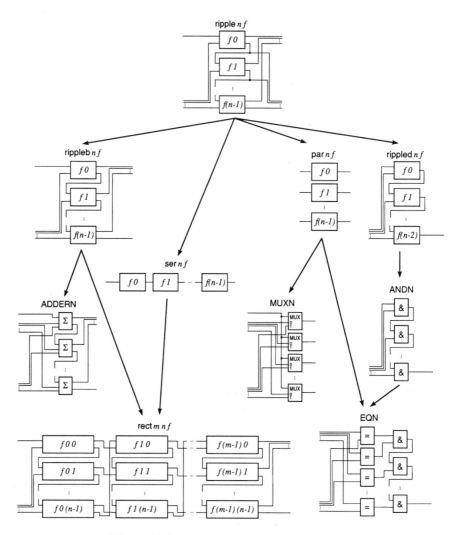

Figure 11.4 Example II of derived structures

via its interface. Especially, the designer can easily switch from one temporal abstraction to another by just replacing the interface pattern and without changing the functional description. In other approaches, where functional and timing aspects are interwoven (VHDL etc.), changing the interface behaviour means skipping the original design and starting from scratch. By the orthogonalized treatment of functional and interface descriptions, reuse of designs is supported in a systematic manner.

Unlike in many other concepts, there is a common basis for the system level, algorithmic and for the rt- and gate level: dfg-terms. The same dfg-terms may be reused at every abstraction level in a different context. The expression (ANDN 7) can be considered as a combinatorial circuit at the rt-level. But it can as well be considered as an algorithmic description: dfg_interface_cycle(ANDN 7, 5) uniquely describes a sequential circuit implementing the algorithmic description (ANDN 7) with a delay (input to output) of 5 clock cycles. To derive the rt-implementation from this expression, high-level synthesis has to be applied.

Besides the nonrecursive dfg-terms, Gropius also supports p-terms, i.e. general computable functions (programs), as starting point for high-level synthesis. Since dfg-terms are a basic part of p-terms, the function (ANDN 7) could as well be used as a part of some p-term - for example in a boolean condition of a while loop.

11.10 UNIFORM COMMUNICATION PROTOCOL AT THE SYSTEM LEVEL

At the system level (Gropius-3) the interface behaviour of all processes is fixed. It realizes a handshake protocol and is inspired by higher order petri nets. An algorithmic process gets one single data package, evaluates its function on it and submits the resulting data package to its successor as soon as the successor process is able to get a new data package. During evaluating the function and waiting for the successor process to accept the resulting data package, the process cannot accept new data packages from its predecessor.

A the system level, only point-to-point connections are allowed, i.e. every output of a process can only be connected to a single input of another process. As a consequence, the processes can be synthesized separately and independently from each other. A special communication synthesis is not necessary. Since the dfg-terms and p-terms, that are used to build algorithmic processes, are only applied in correspondence with the above mentioned protocol, the reuse of components is supported systematically. This is useful, if complex processes must be partitioned into smaller ones. If an implementation of such a sub process already exists, it can be reused for the realization of the complex system.

Another reuse aspect at the system level are k-processes. They are used for combining several algorithmic processes and to manage the communication between them and can be seen as a sort of glue logic that is placed around the algorithmic processes. K-processes are defined once and for all and can thus be reused in different system descriptions.

It is possible to combine several k-processes in order to generate more complex general communication patterns, and it also makes sense to combine them to modify the interface behaviour of algorithmic processes. This is how the user can define a new protocol at the system level. The standard protocol, which has been described above, allows an algorithmic process to get and to deliver only a single data package. However, it would also be possible to first collect some data packages in order to perform a calculation on all of them at the same time. An example is given in the upper part of figure 1.5. This structure of k-processes and simple dfg-circuit descriptions realizes such a protocol: three serial data packages are combined to a single data package, which is then delivered to another process. Of course, this structure can be generalized for n serial data packages. The lower structure of figure 1.5 realizes the other way around.

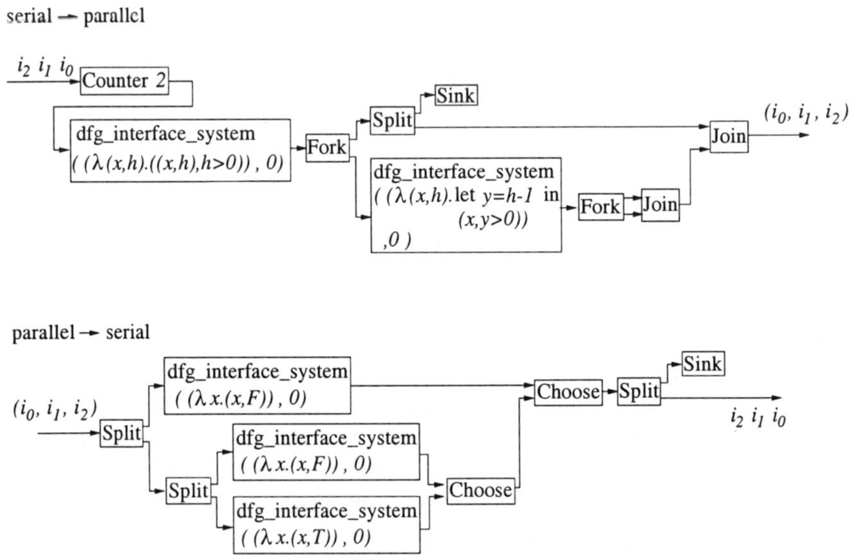

Figure 11.5 Examples of communication schemes

11.11 CONCLUSION

In this paper we have presented the new functional hardware description language Gropius, which supports several powerful concepts within the scope of design reuse. It is a formal, mathematical hardware description language with an exact semantics and allows describing circuits at an abstract, compact, mathematical manner. Furthermore, its expressiveness goes beyond the standard of today's hardware description languages.

We have presented six general features to support the reuse of designs, which could only be introduced very briefly due to lack of space. It is possible to introduce new, user defined functions or even control structures. The polymorphic type system allows defining functions, which can be instantiated in different ways. Gropius supports higher order functions, which means that circuits can be used as parameters of other circuits. Regularity is an important feature in Gropius. Abstract regular circuit descriptions are the starting point to realize totally different components. In contrast to other hardware description languages, Gropius consciously separates functional and temporal aspects in a strict manner. Therefore, algorithmic descriptions can be reused in combination with different interface behaviours without starting from scratch. Finally, a fixed communication protocol at the system level allows reusing whole processes.

Dirk Eisenbiegler received his Diploma (1993) and Ph.D. (1998) degrees in Computer Science from the University of Karlsruhe. From 1993 until 1996 he was research assistant at the department "automation of circuit design" (ACID) at the Forschungszentrum Informatik (FZI) in Karlsruhe. From 1996 until 1998 he worked at the Institute for Circuit Design and Fault Tolerance at the University of Karlsruhe. He is now at the IBM development center in Böblingen, Germany.

Christian Blumenröhr received his Diploma degree in Electrical Engineering at the University of Karlsruhe in 1995. Since 1995 he is a research assistant at the Institute for Computer Design and Fault Tolerance at the same university. There he first was a member of the machine learning group. Since 1996 he is with the formal synthesis group and now works on a research project called "formal circuit design". His research interests include formal methods, high-level synthesis and system level synthesis.

12 LEGAL ASPECTS OF REUSE OF INTELLECTUAL PROPERTY

R. Vogel

Bartsch und Partner
Karlsruhe, Germany

12.1 ISSUES

"Reuse of intellectual property" means that the owner of intellectual property (ideas, concepts, or algorithms that can be implemented in hardware and/or software, in the following referred to as "IP"), which exists, e.g., in the form of software, firmware, or chips, intends to make economic use of this know-how by making it available to a reuser. This transfer makes economic sense for both parties: the owner of the rights receives additional compensation reducing the development costs and increasing his profit. The reuser, on the other hand, does not incur any development costs, since he can resort to previously developed know-how. This means that the reuser's profit also increases. Reused IP technology will probably be used much more extensively in view of increased time-to-market demands and shorter design cycles.

The disadvantage of the reuse of IP for the owner of the rights is that the reuser may use the IP for purposes that are not within the agreed upon scope of use and for which he paid no compensation. If the owner of the rights imposes technical restrictions on the use of the programs (know-how is provided in a "black box"; the reuser is provided with computer programs in machine code and without any development documentation, etc.), the know-how is of less interest to the reuser which reduces his willingness to make payments.

If the owner of the rights provides the reuser with the entire know-how without any technical restrictions, the danger exists that the reuser may use the know-how beyond the agreed upon scope and may compete with the owner of the rights due to his low development costs.

From the legal viewpoint the question arises under which laws, to which extent, and with regard to what kinds of use the owner's know-how is protected (cf. Section 12.2). We need to clarify whether the owner of the rights can protect his know-how by contractual provisions in addition to protection by technical devices and statutory provisions (cf. Section 12.3).

12.2 LEGAL SITUATION

In theory, the legal situation is very simple. In most countries IP is given extensive protection and the owner of the rights can prevent third parties from unauthorized use of his IP or from use which exceeds the scope of the contractual use. As a prerequisite, the IP must have a certain originality. Trivial programs consisting of only a few code lines and which any beginner can write, are not protected. However, European courts interpret originality very broadly. For example, in the past five years there has been no judgment in Germany that refused any software program copyright protection.

Another exception exists, because "Ideas and principles which underlie any element of a computer program, including those which underlie its interfaces" are not protected. Neither German nor European law gives an exact definition of "ideas and principles" and to what extent algorithms are protected by copyright laws and, therefore, these issues are unresolved. However, this problem need not to be discussed in connection with the reuse of IP, since the data contained in the relevant program libraries are complex information protected by copyright law.

12.2.1 Copyright Protection

Within the European Union, software is extensively protected by the Software Directive as of May 14, 1991. In Germany, this protection has been embodied in section 69 a et seq. of the German Copyright Act since the summer of 1993. The owner of the rights has the exclusive right to authorize "the permanent or temporary reproduction, in whole or in part, of a computer program by any means and in any form". "Loading, viewing, running, transmitting or storing a computer program" also requires his prior consent. And finally, the owner of the rights reserves the rights required for a reuse by means of "translation, adaptation, arrangement and any other alteration of a computer program as well as the reproduction of the results of the acts" (see section 69 c par. 1, no. 1, 2, German Copyright Act (UrhG)).

The only exception applicable to the reuse is that an authorized reuser has the right to correct errors in the software and that he may decompile the software to the extent required to obtain interface data (see sections 69 d par. 1, 69 e, German Copyright Act).

By now, all countries within the European economic territory have implemented the Software Directive into national law. In most other countries worldwide software is protected by copyright laws, as for example by the American Copyright Act. During the international GATT trade negotiations, the protection of software was expressly provided for in the TRIPS Agreement (Trade Related Aspects of Intellectual Property Rights) as of December 15, 1993.

12.2.2 Others Kinds of Legal Protection

The legal protection of databases included in the German Copyright Act on January 1, 1998 (sections 87 a et seq. German Copyright Act), based on the EU Database Directive as of March 11, 1996, grants the owner of rights to IP stored in databases extensive rights.

Databases are defined as a "collection of independent works, data, or other material arranged in a systematic or methodical way and individually accessible by electronic or other means". The producer of the database has the exclusive right to reproduce, distribute, and display the database to the public in whole or a material part thereof (section 87 b German Copyright Act). What is important is that not the individual pieces of information contained in the database are protected (these data may be subject to a copyright), but only the database structure and the "electronic material" (thesaurus, index, query language). Also, the database software as such is not protected in connection with the database, but is itself copyrighted.

IP is also protected by other laws, at least indirectly. The patent law deserves special mention. However, in German law, software is not protected as such, because pursuant to § 1 par. 2, no. 3 German Patent Act (PatG) software is not patentable. Software may only be patented indirectly within the framework of hardware/software connections and in the context of the development of new processes (process patent). In other European countries and in the U.S., patent laws and jurisdiction are more generous and grant software more extensive patent protection.

Additional protection is granted by trademark laws and laws against unfair competition. The protection against unfair competition grants subsidiary protection in addition to copyright and database protection. This protection becomes effective, even if the goods or services are not protected as intangible property, when a competitor copies or assumes goods or services and thus achieves a competitive edge with regard to costs or time.

We can, therefore, state the fact that IP is protected comprehensively, in particular by copyright laws. The owner of the rights has the exclusive right to authorize, restrict, or prevent reuse. From the legal viewpoint, the main issue concerning the reuse lies in the definition and verification of the know-how use by contractual and technical means and in proving infringements of these rights.

12.3 CONTRACTUAL AND TECHNICAL REMEDIES

12.3.1 Exact Specification of the Extent of Know-How and Scope of Use

To determine the limits of the reuse an exact definition of the extent and authorized use is required. An exact definition becomes all the more important, since in some jurisdictions the rights granted are only effective if they have been expressly defined (e. g., German and French copyright law). In other legal systems (U.S. copyright system) copyrights can be transferred globally and comprehensively so that in this case the exact definition of the authorized use is required to protect the best interests of the owner of the rights.

If the extent of reuse is specified in detail, it is relatively easy to determine the compensation for the reuse based on use criteria. However, the specification is most important for legal protection: any use which exceeds the specified scope is a copyright infringement. It is easy to prove such a copyright infringement, if the limitations of the authorized use are clearly established.

Together with the exact definition of the released IP information and the authorized kinds of use, the IP owner should include provisions regarding the burden of proof in the license agreements. For example, he may require the reuser to prove compliance with the contractual terms of use by allowing a comparison of the newly written code and the software used as part of the reuse of IP. Provisions which establish the burden of proof must be thoroughly reviewed with regard to their validity in the respective jurisdiction. Under the German Act on General Terms and Conditions, which also protects merchants, the reversal of the burden of proof would be invalid in standard agreements.

12.3.2 Contractual Penalty

It is advisable not to rely merely on legal sanctions in case a person should infringe upon the IP, but to arrange for contractual penalties. If the unauthorized reuser is aware that substantial financial penalties will be imposed in case he exceeds the scope of the authorized use, he will tend to obtain permission for the desired use from the owner of the rights against compensation. Contractual penalties may be combined with a provision shifting the burden of proof to the reuser, so that the contractual penalty becomes due when an infringement is suspected, unless the reuser proves that he has used the IP only to the extent permitted.

12.3.3 Verification and Documentation

In order to facilitate the proof of infringements, verification procedures and documentation obligations should be provided for in the contract. The more precisely the owner of the rights can verify the reuse (e. g., through the obligation to submit software versions, the possibility of computer inspections without prior appointment), the easier it becomes for him to prove a breach. However, this remedy is hardly practical due to the complexity of IP reuse, because otherwise the advantage of this transaction, i.e., the discretionary use of existing information, would be negatively affected.

12.3.4 Technical Measures

In addition to contractual measures, the reuse of IP can and should be verified by technical means (e.g., by means of dongles, encryption technology, run-time licenses) and sanctions should be provided for. From the legal viewpoint, it is advisable that the owner of the rights and reuser expressly agree on technical measures. On the one hand, such a provision acts as a warning, but, on the other hand, the technical restrictions may cause interruptions, which give the reuser a warranty right, unless they were expressly reserved in the contract.

Rupert Vogel received the law degree (1989) from the Heidelberg University, Germany. From 1989 to 1990 he was assistant at the University of Montpellier, France. Since 1993 he is lawyer in the law firm Bartsch und Partner, Karlsruhe, Germany, and lecturer of IT and French law at the University of Mannheim. His professional interests include communications and media, multimedia, computers and software and international law.

13 REFERENCES

[Agst98]   K. Agsteiner, J. Boettger, D. Monjau and S. Schulze. "An Object-Oriented Model for Specification, Prototyping, Implementation and Reuse." *Design Automation and Test in Europe (DATE)*, February 1998.

[Altm94]   J. Altmeyer, S. Ohnsorge and B. Schurmann. "Reuse of Design Objects in CAD Frameworks." *Proceedings of the ICCAD'94*, June 1994.

[Asso94]   Semiconductor Industry Association. *The National Technology Roadmap for Semiconductors*. Semiconductor Industry Association, 1994.

[Asso97]   Semiconductor Industry Association. *The National Technology Roadmap for Semiconductors*. Semiconductor Industry Association, 1997.

[ATM 96]   The ATM Forum Technical Committee. "Traffic Management Specification V4.0." *2af-tm-oo56.000 Letter Ballot*, April 1996.

[Behn98]   B. Behnam, K. Babba, G. Saucier. "IP taxonomy, IP searching in a catalog." *Design Automation and Test in Europe (DATE)*, Designer Track, pages 147–151, February 1998.

[Blum93]   M. Blum et al. "A Workbench for Generation of Components Models." *EURO-DAC'93*, September 1993.

[Blum97]   H. Blumenroth. "Reuse Study for Components of an MPEG-2 decoder (in German)." *Master thesis*, TU Chemnitz, 1997.

[Blum98]   C. Blumenröhr and D. Eisenbiegler. "Performing High-Level Synthesis via Program Transformations within a Theorem Prover." *Digital System Design Workshop at the 24th EUROMICRO 98 Conference*, pages 34 – 37, Västeraas, Sweden, 1998. IEEE-Press.

[Blum99]   C. Blumenröhr. "A Formal Approach to Specify and Synthesize at the System Level." *GI/ITG/GMM Workshop Methoden und Beschreibungssprachen zur Modellierung und Verifikation von Schaltungen und Systemen*, Braunschweig, Germany, 1999.

[Bran97]   H.J. Brand. *CGGEN / Release 1.0 - User Manual*, 1997.

[Chou96]   S.-C. Chou, J.-Y. Chen and C.-G. Chung. "A Behaviour-based Classification and Retrieval Technique for Object-oriented Specification

Reuse." *Software - Practice and Experience*, pages 815–832, July 1996.

[Desi98] Design and Reuse Initiative. "Design and Reuse Initiative" *Technical report, Design and Reuse*, 1998. http://www.design-reuse.com.

[Eise97a] D. Eisenbiegler. "Automata - A Theory Dedicated towards Formal Circuit Synthesis." *Technical Report 14/97*, Universität Karlsruhe, 1997. http://goethe.ira.uka.de/fsynth/publications/postscript/Eise97b.ps.

[Eise97b] D. Eisenbiegler, R. Kumar and C. Blumenröhr. "A Constructive Approach Towards Correctness of Synthesis-Application within Retiming." *European Design & Test Conference*, pages 427–432, Paris, France, March 1997. IEEE Computer Society and ACM/SIGDA, IEEE Computer Society Press.

[Faul98] N. Faulhaber. "Effiziente Aehnlichkeitsmetrik fuer die Wiederverwendung von Intellectual Property." *Diplom Thesis*, Technical University of Karlsruhe, 1998. In German.

[Geis98] J. Geishauser. "StarterKit: A package to speed up IP based simulation in a mixed Verilog - C environment." *IP98 Europe Conference*, October 1998.

[Giam85] N. Giambiasi et al. "An adaptative Evolution Tool for describing General Hierarchical Models, Based on frames and Demons." *Design Automation Conference 1985*, June 1985.

[Grim96] C. Grimm and K. Waldschmidt. "KIR - A graph-based model for description of mixed analog/digital systems." *European Design Automation Conference*, September 1996.

[Grim98a] C. Grimm and K. Waldschmidt. "Repartitioning and technology-mapping of electronic hybrid systems." *Design Automation and Test in Europe (DATE)*, February 1998.

[Grim98b] C. Grimm, F. Heuschen and K. Waldschmidt. "Top-Down Design of Mixed-Signal Systems with KANDIS." *Workshop on System Design Automation (SDA '98)*, March 1998.

[Halb91] N. Halbwachs, P. Caspi, P. Raymond and D. Pilaud. "The synchronous dataflow programming language Lustre." *Proceedings of the IEEE*, 79(9): pages 1305–1320, September 1991.

[ICL97] ICL Manchester. *VHDL+ - Extensions to VHDL for Interface Specification*, 1997.

[IEE96] IEEE. *IEEE Standard VHDL Language Reference Manual Std 1076.3*, 1996.

[Jens92] K. Jensen. *Coloured Petri Nets. Basic Concepts, Analysis Methods and Practical Use. Volume 1, Basic Concepts*. Springer, 1992.

REFERENCES

[Jerr97] A.A. Jerraya, H. Ding, P. Kission and M. Rahmouni. *Behavioral Synthesis and Component Reuse with VHDL.* Kluwer Academic Publishers, 1997.

[Jha 95] P.K. Jha and N.D. Dutt. "Design Reuse through High-Level Library Mapping." *European Design and Test Conference,* pages 345–350, March 1995.

[John91] S.D. Johnson and B. Bose. "DDD: A System for Mechanized Digital Design Derivation." *International Workshop on Formal Methods in VLSI Design,* Miami, Florida, January 1991. ACM/SIGDA. Available as Indiana University Computer Science Department Technical Report No. 323 (rev. 1997).

[Keat98] M. Keating. "A Financial Model for Design Reuse.", September 1998. http://www.synopsys.com/roi/.

[Kiss97] P. Kission, A.A. Jerraya and I. Moussa. "Hardware Reuse." *2nd Workshop on Libraries, Component, Modeling and Quality Assurance,* April 1997. Toledo, Spain.

[Klei98] M. Klein. "Entwurf einer Datenbasis für wiederverwendbare Module." *Diplom Thesis, Eberhard-Karls-Universitaet Tübingen,* 1998. In German.

[Kloo95] C.D. Kloos and P.T. Breuer, editor. *Formal Semantics for VHDL,* volume 307 of *The Kluwer international series in engineering and computer science.* Kluwer, Madrid, Spain, March 1995.

[Koeg97] M. Koegst, P. Conradi, D. Garte and M. Wahl. "Analysis and Classification of Reuse Strategies and Tasks for the Circuit Design." *International Workshop Logic and Architecture Synthesis,* December 1997.

[Koeg98] M. Koegst, S. Ruelke, J. Schoenherr, M. Gulbins, I. Schreiber, E. Fordran and B. Straube. "A Reuse Concept with Verification Aspects and Requirements." *Workshop on System Design Automation (SDA '98),* March 1998.

[Krah96] H. Krahn, J. Eindorf, M. Harnisch, M. Owzar and S. Wolf. "A Single Chip MPEG-2-HDTV Video Decoder (in German)." *3. SICAN-Herbsttagung,* 1996.

[Krue92] C. W. Krueger. "Software Reuse." *ACM Computing Surveys,* 24(2), June 1992.

[Kuma96] R. Kumar, C. Blumenröhr, D. Eisenbiegler and D. Schmid. "Formal Synthesis in Circuit Design-A Classification and Survey." In M. Srivas and A. Camilleri, editors, *Formal Methods in Computer-Aided Design. First International Conference, FMCAD'96,* number 1166 in Lecture

Notes in Computer Science, pages 294–309, Palo Alto, CA, USA, November 1996. Springer-Verlag.

[Lear93] J.O. Leary, M. Linderman, M. Leeser and M. Aagaard. "HML: A Hardware Description Language Based on Standard ML." In D. Agnew, L. Claesen, and R. Camposano, editors, *IFIP Conference on Hardware Description Languages and their Applications*, pages 327–334, Ottawa, Ontario, Canada, April 1993. IFIP WG10.2 International Conference, North-Holland.

[Lebo97] S. Leboch, M. Ryba and U. Baitinger. "Wiederverwendung - Kann der Schaltungsentwurf von der Software-Entwicklung lernen, oder umgekehrt?" *1. Workshop Wiederverwendung im Schaltungsentwurf,* Karlsruhe, Germany, pages 55–64, September 1997. In German.

[Lewi97] J. Lewis. "Intellectual Property Components." Artisan Components Inc., http:\\www.artisan.com, 1997.

[Mart97] F. Martinolle, D. Corlette and S. Pattanam. "A VHDL Procedural Interface for VHDL: VHPI." *IEEE/VIUF International Workshop on Behavioural Modeling and Simulation*, October 1997.

[Mead80] C.A. Mead and L.A. Conway. *Introduction to VLSI Systems*. Addison-Wesley Publishing Company, 1980.

[Meye98] V. Meyer zu Bexte and A. Stuermer. "Design Reuse Experiment for Analog Modules." *2nd Workshop Reuse Techniques for VLSI Design,* Karlsruhe, Germany, pages 81–86, September 1998.

[Morg97] D. Morgan. *Numerical Methods for DSP Systems in C*. John Wiley & Sons, Inc., 1997.

[Mori93] J.D. Morison and A.S. Clarke. *ELLA 2000: A language for electronic system design*. McGraw-Hill, 1993.

[Muel98] S. Mueller and F. Hofmann. "An Easy-to-Use Web-Based Library for HDL Models and IP." *2nd Workshop Reuse Techniques for VLSI Design,* Karlsruhe, Germany, pages 119–126, September 1998.

[Nepp98] F. Neppl. "Viewpoint: Intellectual property will cure IC design woes." *IEEE Spectrum*, pages 26–27, January 1998.

[Oehl98] P. Oehler, I. Vollrath, P. Conradi and R. Bergmann. "Are You READee for IPs?" *2nd Workshop Reuse Techniques for VLSI Design,* Karlsruhe, Germany, pages 1–12, September 1998.

[Olco98] S. Olcoz, L. Ayuda, I. Izaguirre and O. Penalba. "VHDL Teamwork, Organization Units and Workspace Management." *Design Automation and Test in Europe (DATE)*, February 1998.

[Payn98] B. Payne. "Keynote Address IP'98." *IP'98 Europe Conference*, October 1998.

REFERENCES

[Prei95] V. Preis, R. Henftling, M. Schutz and S. Maerz-Roessel. "A Reuse Scenario for the VHDL-Based Hardware Design Flow." *EURO-DAC'95*, September 1995. pages 464-469.

[Rade97] M. Radetzki, W. Putzke-Roeming and W. Nebel. *Objective VHDL: The Object-Oriented Approach to Modeling and Design*. EMMSEC'97, 1997.

[Rows97] J.A. Rowson and A. Sangiovanni-Vincentelli. "Interface-based Design." *31st Design Automation Conference*, 1997.

[Rumb94] J. Rumbaugh et al. *Objektorientiertes Modellieren und Entwerfen*. Carl Hanser and Prentice-Hall, 1994. In German.Translation from: Object-Oriented Modeling and Design, ISBN 0-13-629841-9, Prentice-Hall, 1990.

[Scha98] S. Scharfenberg. "Timing Characterization of Large Modules." *IP98 Europe Conference*, October 1998.

[Schl97] T. Schleinig, H. Krahn, U. Moslehner et. al. "Design of Components for an MPEG-2-HDTV Video Decoder (in German)." *F&M (Feinwerktechnik, Mikrotechnik, Mikroelektronik)*, October 1997.

[Schu98] G. Schumacher and W. Nebel. "Object-Oriented Modelling of Parallel Hardware Systems." *Design Automation and Test in Europe (DATE)*, February 1998.

[Sedn97] M. Sedner. "Redesign of an Inverse Quantiser for an MPEG-2 Decoder (in German)." *Technical Report*, TU Chemnitz, 1997.

[Seep96] R. Seepold, A. Kunzmann and W. Rosenstiel. "Hardware Design Methodology for Efficient Reuse." *2nd World Conference on Integrated Design and Process Technology*, pages 423–430, December 1996.

[Seep98] R. Seepold. *A Hardware Design Methodology with special Emphasis on Reuse and Synthesis*. Berichte aus der Informatik. Shaker Verlag GmbH Aachen, Ph.D. Thesis, University of Tübingen, 1998. ISBN 3-8265-3417-4.

[Sehg94] N.K. Sehgal, C.Y.R. Chen and J.M. Acken. "An Object-Oriented Cell Library Manager." *International Conference on CAD*, pages 750–753, November 1994.

[Shar95] R. Sharp and O. Rasmussen. "The T-Ruby Design System." In *IFIP Conference on Hardware Description Languages and their Applications*, pages 587–596, 1995.

[Sieg97] J. Siegl. "Systemneutrale Verwaltung wiederverwendbarer Schaltkreismodule mittels eines Daten-Management-Systems." *1. Workshop Wiederverwendung im Schaltungsentwurf,* Karlsruhe, Germany, pages 35–40, September 1997. In German.

[Sieg98a]   R. Siegmund. "Untersuchungen zu Methoden des Design Reuse." *Master thesis*, TU Chemnitz, 1998. In German.

[Sieg98b]   R. Siegmund, D. Mueller, H. v. Sychowski and J. Lancaster. "Specification and verification of complex digital systems using VHDL+." *IP 98 Europe Confernce*, October 1998.

[Stue98]   A. Stuermer and V. Meyer zu Bexten. "Design Information System with Documentation Access for Analog Circuits." *First Workshop MEDEA A409 (SADE)*, The Hague, The Netherlands, September 1998.

[Swam95]   S. Swamy, A. Molin and B. Covnot. "OO-VHDL: OO-VHDL Object-Oriented Extensions to VHDL." *IEEE Transactions on Computer Science*, pages 18–26, 1995.

[Syno94]   Inc. Synopsys. *SYNOPSYS, Design-Ware Manuals*, June 1994.

[Trim81]   S. Trimberger, J.A. Rowson, C.R. Lang and J.P. Gray. "A Structural design methodology and associated software tools." *IEEE Transactions on Circuits and Systems*, 28, July 1981.

[VSIA97]   VSIA Virtual Socket Interface Alliance. "VSI Alliance Architecture Document.", 1997, http://www.vsi.org/library/vsi-or.pdf.

[Wage95]   D. Wagenblasst and W. Thronicke. "An Approach for Classification of Integrated Circuits by a Knowledge Conserving Library Concept." *European Design Automation Conference*, pages 40–45, September 1995.

INDEX

A
ALF 46
analog 92, 97, 103, 107
applet 32
ATM 9
attribute 22, 29, 35

B
behaviour 92
Boundary Scan 72
business model 38, 44
business unit 43

C
CA 24
CCITT 122
CE 24
CFI 2
Characteristic Vector see CV
Charateristic Attribute see CA
classification scheme 25, 32
classification technique 22
coding guideline 57
communication 65
Component Environment see CE
component generation 115
conceptual distance 24
conceptual similarity 25
contractual penalty 142
copyright 140
copyright protection 140
core 39
    design of reusable 46
    functionality 39
    integration phase 42
    library 47
    quality 39, 45
    realization phase 42
    supply process 40, 41
    support and maintenance 39
co-simulation 58
CSP see core supply process
CV 25
    Function 29, 33
    Input 29
    Output 29

D
database 31, 50
dataflow graphs 94
DATE 3
design
    cost 9
    flow 45, 91, 104
    instance 53
    modular 14
    parametrisation 115
    reuse of 2
    rule 14, 83
    support 122
DesignObjects 50
documentation 51, 122
DREAM 103
DSP 39

E
ECSI 1, 2
EDA Roadmap 3
EDAC 3
EDIF 2

EUREKA 6

F

financial model 42
firmware 139
floorplanning 14, 73
formal synthesis 125
Full Scan 72
functional verification 114

G

GATT 141
generic model 10
generic module 94

H

hardware description 126
hardware/software co-design 7
high-level synthesis 13

I

Intellectual Property see IP
interface description 133
Internet 31, 36
Intranet 31
IP 21, 38, 49, 83, 94, 114, 139, 141
    business 49
    development 50
    documentation 51
    exchange 4, 44
    generic 113
    hard 63, 73
    matching 30
    meta data 76
    protection 64
    provider 44, 56
    similar 30
    soft 63
    standardisation 2
    vendor 49

J

JESSI 6

K

keyword 22, 35

L

legal protection 141
legal situation 140
LIBerty 46
library concept 80

M

MATLAB 82
MEDEA 1, 6
mixed signal 91
module generator 94
module mapping 94
MPEG 58, 112

O

Objective VHDL 116
object-oriented
    data model 24
    modelling 115
OLA 46
OMF 3
OMI 46

P

PathMill 68
PCI 39
PDDP 40
Perl/Tk 54
placement 73
polymorphism 131
power estimation 71
protection 140

Q

quality 84

R

relational database 91
release management 56

reusability 117
reusable
    blocks 15
    code 53
    component 25
reuse
    analog 103
    database 22
    design by 112
    design for 13, 14, 113
    in-house 61
    library 80
    macro-block 10
    methodology 49, 81
    of design 112
Reuse Management System see RMS
RMS 22
    Classification 28, 35
    database 32
    metric 25
    similarity metric 26
    Taxonomy 27
routing 73
royalty 44

S

SADE 109
Scan Test see Full Scan
search criteria 104
search engine 105
Self Test 72
SIA roadmap 8, 37
similarity metric 22, 24, 35
simulation model 64
SISC 10
SoC 4, 38
soft core 50
software 139
SPICE 68, 100, 106
style guide 83
SuperVISE 121
synthesis 130
system partitioning 12
System-on-chip see SoC

T

taxonomy 22, 35
taxonomy leaf 29
technology tables 98
test strategy 72
testbench 53, 60, 114
timing analysis 68
timing characterization 67
TRIPS 141

U

USB 39

V

VC 22, 114
VCA 24
Vector of Characteristic Attributes see VCA
Verification 54
verification 46, 54, 114, 122
Verilog 56, 64
VHDL 2, 17, 56, 82, 126
VHDL+ 121
Virtual Component see VC
Virtual Socket Interface see VSI
VSI 1, 4, 46, 47, 61
VSIA see VSI